FUNCTIONAL POLYMERIC COMPOSITES

Macro to Nanoscales

FUNCTIONAL POLYMERIC COMPOSITES

Macro to Nanoscales

Edited by

**Chin Hua Chia, PhD
Chin Han Chan, PhD
Sabu Thomas, PhD**

Apple Academic Press Inc.
3333 Mistwell Crescent
Oakville, ON L6L 0A2 Canada

Apple Academic Press Inc.
9 Spinnaker Way
Waretown, NJ 08758 USA

© 2018 by Apple Academic Press, Inc.
No claim to original U.S. Government works
Printed in the United States of America on acid-free paper
International Standard Book Number-13: 978-1-77188-499-0 (Hardcover)
International Standard Book Number-13: 978-1-315-20745-2 (eBook)

All rights reserved. No part of this work may be reprinted or reproduced or utilized in any form or by any electronic, mechanical or other means, now known or hereafter invented, including photocopying and recording, or in any information storage or retrieval system, without permission in writing from the publisher or its distributor, except in the case of brief excerpts or quotations for use in reviews or critical articles.

This book contains information obtained from authentic and highly regarded sources. Reprinted material is quoted with permission and sources are indicated. Copyright for individual articles remains with the authors as indicated. A wide variety of references are listed. Reasonable efforts have been made to publish reliable data and information, but the authors, editors, and the publisher cannot assume responsibility for the validity of all materials or the consequences of their use. The authors, editors, and the publisher have attempted to trace the copyright holders of all material reproduced in this publication and apologize to copyright holders if permission to publish in this form has not been obtained. If any copyright material has not been acknowledged, please write and let us know so we may rectify in any future reprint.

Trademark Notice: Registered trademark of products or corporate names are used only for explanation and identification without intent to infringe.

Library and Archives Canada Cataloguing in Publication

Functional polymeric composites : macro to nanoscales / edited by Chin Hua Chia, PhD, Chin Han Chan, PhD, Sabu Thomas, PhD.
Includes bibliographical references and index.
Issued in print and electronic formats.
ISBN 978-1-77188-499-0 (hardcover).--ISBN 978-1-315-20745-2 (PDF)
1. Polymeric composites. 2. Polymers. 3. Nanocomposites (Materials).
I. Chia, Chin Hua (PhD), editor II. Chan, Chin Han, editor III. Thomas, Sabu, editor

TA455.P58F86 2017	620.1'92	C2017-904606-3	C2017-904607-1

Library of Congress Cataloging-in-Publication Data

Names: Chia, Chin Hua, editor.
Title: Functional polymeric composites : macro to nanoscales / editors, Chin Hua Chia, PhD, Chin Han Chan, PhD, Sabu Thomas, PhD.
Description: Toronto ; New Jersey : Apple Academic Press, 2015. | Includes bibliographical references and index.
Identifiers: LCCN 2017030565 (print) | LCCN 2017033154 (ebook) | ISBN 9781315207452 (ebook) | ISBN 9781771884990 (hardcover : alk. paper)
Subjects: LCSH: Polymeric composites.
Classification: LCC TA418.9.C6 (ebook) | LCC TA418.9.C6 F868 2015 (print) | DDC 620.1/92-dc23
LC record available at https://lccn.loc.gov/2017030565

Apple Academic Press also publishes its books in a variety of electronic formats. Some content that appears in print may not be available in electronic format. For information about Apple Academic Press products, visit our website at **www.appleacademicpress.com** and the CRC Press website at **www.crcpress.com**

ABOUT THE EDITORS

Chin Hua Chia, PhD

Chin Hua Chia is an Associate Professor in the Materials Science Programme, School of Applied Physics, Universiti Kebangsaan Malaysia (UKM) (also known as National University of Malaysia). He obtained his PhD in Materials Science (UKM, Malaysia) in 2007. His core research interests include developing polymer nanocomposites, bio-polymers, magnetic nanomaterials, and bio-adsorbents for wastewater treatment. He has published more than 120 research articles and more than 80 publications in conference proceedings. He is an Editor of *Sains Malaysiana* and the Editor-in-Chief of the *Polymers Research Journal* (Nova Science Publisher). He has received the Best Young Scientist Award (2012) and the Excellent Service Award (2013) from UKM, and the Young Scientist Award (2014) from the Malaysian Solid State Science and Technology Society (MASS). He has recently received the Distinguished Lectureship Award (2017) from the Chemical Society of Japan (CSJ).

Chin Han Chan, PhD

Chan Chin Han is an Associate Professor at Universiti Teknologi MARA, Malaysia. She was appointed as Visiting Scientist at China University of Petroleum, Beijing, China (2011–2012), as Chair Professor on Advances in Hybrid Materials at Mahatma Gandhi University, Kottayam, India (2014), and National Representative (Malaysia) of the International Union of Pure and Applied Chemistry (IUPAC) for the Polymer Division (2014–2017). She has been the co-chief editor of *Materials Mind* (a quarterly magazine of the Institute of Materials, Malaysia) and one of the editors of several books published by the Royal Society of Chemistry (2013) and Apple Academic Press (2014). Her research interest is devoted to physical properties of epoxidized natural rubber-based blends and thermoplastic elastomer, biodegradable polyester/polyether blends, and solid polymer electrolytes. She has published books and many articles in professional journals, contributed chapters in books, and presented invited talks at many professional and academic conferences.

Sabu Thomas, PhD

Professor Sabu Thomas has more than 30 years' experience in polymer science and technology and has contributed greatly to the research and development of nanoscience and nanotechnology. He is presently the Director of the International and Inter University Centre for Nanoscience and Nanotechnology and full Professor of Polymer Science and Engineering at the School of Chemical Sciences of Mahatma Gandhi University, Kottayam, Kerala, India. He is one of the pioneers of the field of polymer science and technology and has published over 780 peer-reviewed research papers, reviews, and book chapters. He has co-edited nearly 73 books and is the inventor of five patents. His H-index is 80M with nearly 30245 citations to date. In addition, he has delivered over 310 plenary/inaugural and invited lectures at national/international meetings over 30 countries. He has supervised 79 PhD theses. He is currently the chief editor of *Nano-Structures & Nano-Objects* and also serves the reviewer for several international journals. He has established a state-of-the-art laboratory at Mahatma Gandhi University in the area of polymer science and engineering and nanoscience and nanotechnology.

Professor Thomas has received a number of national and international awards, including a Fellowship of the Royal Society of Chemistry, London; MRSI, Nano Tech, and CRSI medals; a Distinguished Faculty Award; and the Sukumar Maithy Award for being the best polymer researcher in the country. He is on the list of most productive researchers in India and holds a position of the No. 5 spot. He has received the Honoris Causa (DSc) by the University of South Brittany, Lorient, France, in 2015, and in May 2016 he was awarded the Loyalty Award at the International Materials Technology Conference. Recently in 2017 he has been selected to receive a DSc from the University Lorrainem France. To add to this, he recently was also selected for a life-time achievement award by the Indian Nano-Biologist Association (INBA).

Prof Sabu Thomas received his PhD from IIT, Kharagpur, and then joined as a Senior Visiting Researcher at Katholieke University, Leuven, Belgium, and Laval University, Quebec, Canada. Subsequently, he served

About the Editors vii

as Associate Professor and Professor at Mahatma Gandhi University, Kottayam, Kerala, India.

Dr. Thomas's ground-breaking inventions in polymer nanocomposites, polymer blends, green nanotechnological, and nano-biomedical sciences have made transformative differences in the development of new materials for the automotive, space, housing, and biomedical fields.

CONTENTS

List of Contributors .... *xi*

List of Abbreviations .... *xiii*

Preface .... *xvii*

1. **A Review on Mechanical Properties of Semicrystalline/Amorphous Polymer Blends** 1

   Srinivasarao Yaragalla, Nithin Chandran, Chin Han Chan, Nandakumar Kalarikkal, and Sabu Thomas

2. **Preparation of Polysaccharide-Based Composite Materials Using Ionic Liquids** 33

   Jun-Ichi Kadokawa

3. **Microcrystalline Cellulose: An Overview** 55

   Srimanta Sarkar, Celine Valeria Liew, Josephine Lay Peng Soh, Paul Wan Sia Heng, and Tin Wui Wong

4. **Bio-Based Phenol Formaldehyde from Lignocellulosic Biomass** 75

   Rasidi Roslan, Sarani Zakaria, Chin Hua Chia, Umar Adli Amran, and Sharifah Nabihah Syed Jaafar

5. **Impedance Spectroscopy: A Practical Guide to Evaluate Parameters of a Nyquist Plot for Solid Polymer Electrolyte Applications** 97

   Siti Rozana Bt. Abdul Karim, Chin Han Chan, and Lai Har Sim

6. **Polymerization of Natural Oils for a Quartz Crystal Microbalance-Based Gas Sensor Application** 129

   Rashmita Das, Panchanan Pramanik, and Rajib Bandyopadhyay

7. **Emergence of a New Nanomaterial: Nanocellulose and Its Nanocomposites** 163

   Chi Hoong Chan, Chin Hua Chia, and Sarani Zakaria

8. **Qualitative Fourier Transform Infrared Spectroscopic Analysis of Polyether-Based Polymer Electrolytes** 197

   Siti Rozana Bt. Abdul Karim and Chin Han Chan

Index 243

LIST OF CONTRIBUTORS

Umar Adli Amran
Bioresouces and Biorefinery Laboratory, Faculty of Science and Technology, Universiti Kebangsaan Malaysia, Bangi 43600, Selangor, Malaysia

Rajib Bandyopadhyay
Department of Instrumentation and Electronics Engineering, Jadavpur University, Kolkata 700098, West Bengal, India. E-mail: bandyopadhyay.rajib@gmail.com

Chin Han Chan
Faculty of Applied Sciences, Universiti Teknologi MARA, Shah Alam 40450, Malaysia. E-mail: cchan@salam.uitm.edu.my, cchan_25@yahoo.com.sg

Chi Hoong Chan
School of Applied Physics, Faculty of Science and Technology, Universiti Kebangsaan Malaysia, Bangi 43600, Selangor, Malaysia
Bioresources and Biorefinery Laboratory, School of Applied Physics, Faculty of Science and Technology, Universiti Kebangsaan Malaysia, Bangi 43600, Selangor, Malaysia. E-mail: dinotim88@gmail.com

Nithin Chandran
School of Chemical Sciences, Mahatma Gandhi University, Kottayam 686560, Kerala, India
Faculty of Applied Sciences, Universiti Teknologi MARA, Shah Alam 40450, Malaysia

Chin Hua Chia
Bioresouces and Biorefinery Laboratory, Faculty of Science and Technology, Universiti Kebangsaan Malaysia, Bangi 43600, Selangor, Malaysia. E-mail chiachinhua@yahoo.com / chia@ukm.edu.my

Rashmita Das
Department of Instrumentation and Electronics Engineering, Jadavpur University, Kolkata 700098, West Bengal, India. E-mail: kunurashmita@gmail.com

Paul Wan Sia Heng
GEA-NUS Pharmaceutical Processing Research Laboratory, Department of Pharmacy, Faculty of Science, National University of Singapore, 18 Science Drive 4, Singapore 117543, Singapore

Sharifah Nabihah Syed Jaafar
Bioresouces and Biorefinery Laboratory, Faculty of Science and Technology, Universiti Kebangsaan Malaysia, Bangi 43600, Selangor, Malaysia
Faculty of Applied Sciences, Universiti Teknologi MARA, Shah Alam 40450, Malaysia

Jun-ichi Kadokawa
Graduate School of Science and Engineering, Kagoshima University, 1-21-40 Korimoto, Kagoshima 890-0065, Japan

Nandakumar Kalarikkal
School of Pure and Applied Physics, Mahatma Gandhi University, Kottayam 686560, Kerala, India
International and Inter University Center for Nanoscience and Nanotechnology, Mahatma Gandhi University, Kottayam 686560, Kerala, India

Siti Rozana Bt. Abdul Karim
Faculty of Applied Sciences, Universiti Teknologi MARA, Shah Alam 40450, Malaysia. E-mail: sitirozanaabdkarim@gmail.com

Celine Valeria Liew
GEA-NUS Pharmaceutical Processing Research Laboratory, Department of Pharmacy, Faculty of Science, National University of Singapore, 18 Science Drive 4, Singapore 117543, Singapore. E-mail: phalcv@nus.edu.sg

Panchanan Pramanik
Department of Basic Science, MCKV Institute of Engineering, Liluah, Howrah 711204, West Bengal, India. E-mail: pramanik1946@gmail.com

Rasidi Roslan
Material Technology Department, Faculty of Industrial Sciences and Technology, Universiti Malaysia Pahang, Lebuhraya Tun Razak, 26300 Gambang Kuantan, Pahang, Malaysia. E-mail: rasidi@ump.edu.my

Srimanta Sarkar
GEA-NUS Pharmaceutical Processing Research Laboratory, Department of Pharmacy, Faculty of Science, National University of Singapore, 18 Science Drive 4, Singapore 117543, Singapore Pharma Tech Ops Manufacturing Science and Technology, Novartis Pharma AG, Basel, Switzerland

Lai Har Sim
Centre of Foundation Studies, Universiti Teknologi MARA, Puncak Alam 42300, Malaysia. E-mail: marialhsim@salam.uitm.edu.my

Josephine Lay Peng Soh
GEA-NUS Pharmaceutical Processing Research Laboratory, Department of Pharmacy, Faculty of Science, National University of Singapore, 18 Science Drive 4, Singapore 117543, Singapore Pharma Tech Ops Manufacturing Science and Technology, Novartis Pharma AG, Basel, Switzerland

Sabu Thomas
Centre for Nanoscience and Nanotechnology, Mahatma Gandhi University, Kottayam, Kerala, India
School of Chemical Sciences, Mahatma Gandhi University, Kottayam 686560, Kerala, India
Faculty of Applied Sciences, Universiti Teknologi MARA, Shah Alam 40450, Malaysia

Tin Wui Wong
Non-Destructive Biomedical and Pharmaceutical Research Centre, Universiti Teknologi MARA, Puncak Alam 42300, Selangor, Malaysia
Particle Design Research Group, Faculty of Pharmacy, Universiti Teknologi MARA, Puncak Alam 42300, Selangor, Malaysia
CoRe Frontier Materials and Industry Application, Universiti Teknologi MARA, Shah Alam 40450, Selangor, Malaysia. E-mail: wongtinwui@salam.uitm.edu.my

Srinivasarao Yaragalla
International and Inter University Center for Nanoscience and Nanotechnology, Mahatma Gandhi University, Kottayam 686560, Kerala, India
Faculty of Applied Sciences, Universiti Teknologi MARA, Shah Alam 40450, Malaysia

Sarani Zakaria
Bioresouces and Biorefinery Laboratory, Faculty of Science and Technology, Universiti Kebangsaan Malaysia, Bangi 43600, Selangor, Malaysia. E-mail: szakaria@ukm.edu.my, sarani_zakaria@yahoo.com

LIST OF ABBREVIATIONS

ABS	acrylonitrile-butadiene-styrene
AcMIMBr	1-(4-acryloyloxybutyl)-3-methylimidazolium bromide
AcVIMBr	1-(4-acryloyloxypropyl)-3-vinylimidazolium bromide
AFM	atomic force microscope
AMIMBr	1-allyl-3-methylimidazolium bromide
ATR	attenuated total reflectance
BC	bacterial cellulose
BMIMCl	1-butyl-3-methylimidazolium chloride
BPE	bio-phenolic elastomer
BTAI	blocked allyl (3-isocyanate-4-tolyl) carbamate
CL	caprolactone
CNC	cellulose nanocrystals
CNF	cellulose nanofibrils
CNT	carbon nanotubes
CPE	constant phase element
CR	chloroprene rubber
CrI	crystallinity index
DCP	dicumyl peroxide
DHMP	dihydroxymethyl phenol
DMA	dynamic mechanical analysis
DP	degree of polymerization
DSC	differential scanning calorimetry
ENR	epoxidized natural rubber
EPDM	ethylene-propylene-diene monomer
EPFBF	empty palm fruit bunch fibers
EPR	ethylene-propylene rubber
FRGS	fundamental research grant scheme
FTIR	Fourier transform infrared
FTIR	Fourier-transform infrared
GMA	glycidyl methacrylate
GPE	gel polymer electrolyte
H_2SO_4	sulphuric acid
H_3PO4	phosphoric acid

HBr	hydrobromic acid
Hcl	hydrochloric acid
HDPE	high density poly ethylene
HMP	hydroxymethyl phenols
HPC	hydroxypropylcellulose
HRJ	hydroxyl methyl phenolic resin
IB	internal bonding
ICH	International Conference on Harmonisation
IL	ionic liquids
IPA	isopropyl alcohol
IPN	interpenetrating polymer network
iPS	isotactic poly styrene
IS	impedance spectroscopy
LA	l-lactide
LCST	lower critical solution temperature
LDPE	latex/low density polyethylene
LPE	liquid polymer electrolyte
MAH-g	maleic anhydride grafted
MB	methylene blue
MCC	microcrystalline cellulose
MCF	cellulose microfibrils
MEMS	microelectromechanical systems
MFC	microfibrillated cellulose
MMA	methylmethacrylate
MOHE	ministry of higher education
MOR	modulus of rupture
$NaClO_2$	sodium chlorite
NaOH	sodium hydroxide
NBR	nitrile-butadiene rubber
NF	national formulary
NFC	nanofibrillated cellulose
NR	natural rubber
O	oxygen
OE	oligoesters
PA	poly amide
PAN	polyacrylonitrile
PBMA	poly butyl methacrylate
PBT	poly butylene terephthalate

List of Abbreviations

PBT	poly butylene terephthalate
PBzMA	poly benzyl methacrylate
PC	poly carbonate
PCL	poly ε-caprolactone
PCMA	poly cyclohexyl methacrylate
PE	poly ethylene
PEFB	palm empty fruit bunch fibers
PEO	poly ethylene oxide
PEO	polymer electrolyte
PER	polyester resins
PET	poly ethylene terephthalate
PET	poly ethylene terephthalate
PF	phenol formaldehyde
PHA	polyhydroxyalkanoate
PiBMA	poly iso-butylmmethacrylate
PIL	polymeric Ionic liquid
PLA	poly lactic acid
PMMA	poly methyl methacrylate
PMVE-MAc	poly methyl vinyl ether-maleic acid
PnBMA	poly n-butyl methacrylate
PP	poly propylene
PP	polypropylen
PPE	poly phenylene ether
PPhMA	poly phenyl methacrylate
PPMA	poly propyl methacrylate
PPO	poly phenylene oxide
PRISM	statistical mechanical polymer reference interaction site model
PS	poly styrene
PtBMA	poly tert-butyl methacrylate
PVA	poly vinyl alcohol
PVC	poly vinyl chloride
PVDF	poly vinylidene fluoride
PVdF	polyvinylidene fluoride
PVP	poly vinyl pyrrolidone
PVPh-HEM	poly 4-vinylphenol-*co*-2-hydroxyethyl methacrylate
QCM	quartz crystal microbalance
QSPR	quantitative structure property relationship

SAGMA	styrene-acrylonitrile glycidyl methacrylate
SAN	styrene acrylonitrile
SBR	styrene-butadiene rubber
SEBS	styrene-b-(ethylene-co-1-butene)-b-styrene
SEM	scanning electron microscopic
SI-ATRP	surface-initiated atom transfer radical polymerization
SP	dimethyl phenolic resin
SPE	solid polymer electrolyte
SPE	solid polymer electrolyte
TEM	transmission electron microscope
TEMPO	2,2,6,6,-tetramethylpiperidine
TGA	thermal gravimetric analysis
THMP	trihydroxymethyl phenol
TOCN	TEMPO-oxidized cellulose nanofibrils
TOR	trans-polyoctylene rubber
UF	urea formaldehyde
VOC	volatile organic compound
VVBnIMCl	1-vinyl-3-(4-vinylbenzyl)imidazolium chloride
XRD	x-ray diffraction
ZnSe	zinc selenide

PREFACE

Research in advanced polymers and polymer composites has been playing a very important role in the development of science and technology. It is an interdisciplinary branch of science that involves various disciplines, including physics, chemistry, biology, mathematics, engineering, etc. During the International Symposium on Advanced Polymeric Materials (ISAPM, 2012) under the auspices of the eighth Materials and Technology Conference and Exhibition (IMTCE, 2012) organized in Malaysia in 2012, we were able to bring together researchers from universities and industries from different countries. After the symposium, a book titled *Physical Chemistry and Macromolecules: Macro to Nanoscales,* which is comprised of chapters contributed by invited speakers of the symposium, had been published. The chapters focus on applications emphasizing the latest trends of advanced polymeric materials, such as polymer blends, micro- to nano-composites, and biopolymers.

In 2014, we again organized the same symposium under the auspices of the ninth International Materials Technology Conference and Exhibition (IMTCE, 2014) in Kuala Lumpur, Malaysia. More than 150 participants from different countries attended and presented their latest research findings at the symposium. Some of the invited speakers have also contributed chapters to this book. This book offers a brief review of the state-of-the-art research and provides the latest development on advanced polymers and composites as well as their applications, including polymer blends, preparation of bio-based composite using ionic liquid, preparation of micro- and nano-crystalline cellulose, bio-based phenolic resin, etc.

Finally, we would like to thank all the contributing authors for their efforts to make this book possible. We would also like to acknowledge the referees for their constructive comments to improve the quality of the book.

—**Chin Hua Chia, PhD**
Chin Han Chan, PhD
Sabu Thomas, PhD

CHAPTER 1

A REVIEW ON MECHANICAL PROPERTIES OF SEMICRYSTALLINE/ AMORPHOUS POLYMER BLENDS

SRINIVASARAO YARAGALLA[1,3], NITHIN CHANDRAN[2,3], CHIN HAN CHAN[3,*], NANDAKUMAR KALARIKKAL[1,4], and SABU THOMAS[2,3]

[1]*International and Inter University Center for Nanoscience and Nanotechnology, Mahatma Gandhi University, Kottayam 686560, Kerala, India*

[2]*School of Chemical Sciences, Mahatma Gandhi University, Kottayam 686560, Kerala, India*

[3]*Faculty of Applied Sciences, Universiti Teknologi MARA, Shah Alam 40450, Malaysia*

[4]*School of Pure and Applied Physics, Mahatma Gandhi University, Kottayam 686560, Kerala, India*

*Corresponding author. E-mail: cchan@salam.uitm.edu.my; cchan_25@yahoo.com.sg

CONTENTS

Abstract 3
1.1   Introduction 3
1.2   Classification of Polymers 6
1.3   Classification of Multi-Component Polymeric Systems 8
1.4   Mechanical Properties 9

1.5 Mechanical Properties of Semicrystalline/Amorphous
    Binary Blends 9
1.6 Conclusion and Future Perspectives 24
Keywords 25
References..25

ABSTRACT

Fundamental classification of polymers and multicomponent polymeric systems was discussed in this chapter initially. The core part of this chapter encompasses the mechanical properties, mainly tensile strength, Young's modulus, impact strength, and elongation at break of semicrystalline-amorphous polymer blends with respect to the polymer blend ratio, morphology, chemical and physical interactions.

1.1 INTRODUCTION

Most of the useful commercial products are made of multi-component polymeric systems, e.g., parts of computers, automotive parts, etc. Multi-component polymeric systems include compositions which contain more than one component. One of the components should be a polymer while others can be either organic and/or inorganic materials. This chapter enumerates the fundamental aspects of multi-component polymer systems, including binary polymer blends, one polymer with organic or inorganic filler (polymer composites), and binary polymer blends with organic or inorganic fillers (polymer blend composites). Many studies have been carried out on polymer blends as well as polymer or polymer blend composites, with special focus on their morphologies and mechanical properties.[1–10] The morphologies and the intermolecular interactions of components for multi-component systems govern the mechanical, transport, and barrier properties, etc.

In this chapter, our discussion related to polymer blends is limited to semicrystalline/amorphous binary systems. In general, the miscibility of binary polymer blend systems can be defined as the mixing of two components at molecular level which results in homogeneous systems, and they normally exhibit single and compositional-dependent glass transition temperature (T_g). According to thermodynamic point of view, miscibility of two polymers can be defined as the mixing of two polymers which results in positive entropy change (ΔS), negative enthalpy (ΔH), and negative free energy (ΔG). Then, the two polymers are considered to be miscible. The blends consist of poly(styrene) (PS)/poly(phenylene oxide) (PPO) that are miscible, which show miscibility over a wide temperature range. It exhibits single and compositional-dependant T_g of both components containing the blend. Besides, poly(β-hydroxybutyrate)/

poly(p-vinylphenol), poly(ethylene oxide) (PEO)/poly(methyl methacrylate) (PMMA), and isotactic poly(styrene) (iPS)/PS are the other examples of miscible polymer blends.[11–13] In some cases, miscibility of blends may be due to common specific intermolecular interactions between two different polymer chains, e.g., hydrogen bond and dipole–dipole interactions. In general, it is not common to have miscible polymer blends owing to the high molecular weight of polymers. For commercial applications, most of the useful products are immiscible polymer blends rather than miscible polymer blends. The prerequisites for the commercial application of immiscible polymer blends are stable morphologies with reproducible properties. In other words, since the material performance depends not only on the ingredients and their concentration, but also on morphology, either it must be stable and unchanged by the processing conditions or the change must be reproducible. Table 1.1 displays applications and utility of immiscible polymer blends in various fields such as automobile, aerospace, and electronics. Semicrystalline/amorphous blends are the most extensively studied among the three categories of binary polymer blends, i.e., amorphous/amorphous, semicrystalline/amorphous, and semicrystalline/semicrystalline blends.[14–17] The performance of semicrystalline/amorphous blends depends very much on the properties of individual component, phase behavior, and the blend morphology. For example, undesired properties of the products like low impact strength, poor weld lines of injection molded parts, etc., may be the consequences of inadequate dispersion of the minor components in the matrix or poor stabilization morphology of the blends.[18,19] Nevertheless, optimization of the semicrystalline/amorphous blends performance can be done through the

TABLE 1.1 Applications of Immiscible Polymer Blends.[22]

Immiscible Polymer Blends	Applications
PP/EPDM or PP/EPR	Weather-stripping, air dams
PC/ABS	Rocker panel, side door (coupe)
PC/PBT	Rear bumper cover
SAN/EPDM	Construction products and sporting goods.
PA/PPE	Front fender, rear quarter panel

Note: PP, poly(propylene); PC, poly(carbonate); EPR, ethylene-propylene rubber; ABS, acrylonitrile-butadiene-styrene; PBT, poly (butylene terephthalate); SAN, styrene acrylonitrile; EPDM, ethylene-propylene-diene monomer; PA, poly(amide); PPE, poly(phenylene ether).

control of blend morphology *via* molecular structures of the components, blend compositions, processing conditions,[20,21] and so on.

For immiscible blends, they can be categorized as compatible blends and incompatible blends. In the case of incompatible blends, there are macro-phase separations of the components which lead to coarse phase morphology, and hence interface and interfacial adhesions are very poor. These kinds of blends may not be useful due to low mechanical properties without suitable compatibilizer. (PA)/(ABS),[11] poly(ethylene) (PE)/(PC), and (PP)/natural rubber (NR) are common examples of incompatible blends.[23,24] The addition of a suitable compatibilizer to incompatible blends can improve the mechanical properties[25–27] through enhancement on the adhesion between the two phases at the interface and thus, reduce the interfacial tension. In other words, compatible blends show better interface adhesion, which leads to the increase in compatibility of the components at the micro scale, and hence, enhancement in the mechanical properties can be noted. The blends of poly(vinyl alcohol) (PVA)/poly(vinyl pyrrolidone) (PVP), poly(vinylidene fluoride) (PVDF)/PMMA, and poly(ε-caprolactone) (PCL)/poly(styrene-*co*-acrylonitrile) are general examples of compatible blends.[28–30]

PP/NR blends exhibit droplet morphologies where PP is the continuous phase; if it is in excess, then, NR becomes the dispersed phase. On the other hand, if NR is in excess, the continuous phase will be NR and the disperse phase is PP. In general, the morphology of the blends can also be one of the factors in regulating the mechanical properties. Herein, the mechanical properties of PP/NR blends can be regulated by the blend composition as well as the control of the droplet size of the dispersed phase. Generally, the tensile strength and Young's modulus decrease with increasing the NR content in these blends. However, the elongation at break enhances with increasing content of NR.[31,32]

In general, morphologies of polymer composites may be affected by the weight ratio, dimensions, and dispersion of filler in the polymer matrix. In the case of polymer–clay composites, the morphologies of the systems, such as agglomeration of clay in polymer matrix, intercalation, and exfoliation of nanoclay, depend upon the dispersion of clay in the polymer matrix and processing methods.[33,34] The mechanical properties of the polymer composites can be fine-tuned by the dispersion and orientation of filler in the polymer matrix.[35,36]

In the case of polymer blend with filler (polymer blend composites), morphologies of the systems may be affected by the molecular structures of the components, blend composition, processing condition, dispersion and orientation of the filler, etc. There are studies which demonstrate that addition of filler to polymer blends may affect the domain sizes of the respective components.[37,38] For example, consider the droplet morphologies of PP/PS blends in which the amorphous polymer (PS) is the dispersed phase and the semicrystalline polymer (PP) is the continuous phase. The addition of clay to these blends decreases the droplet size of the amorphous phase,[39] and thereby improves the mechanical properties of PP/PS systems as a result of enhanced compatibility between PP and PS induced by the addition of filler.[3,4,40] In some cases, filler addition results in enhanced compatibility between phases in polymers, where the filler may act as a physical cross-linker between polymers.

The scope of discussion of this chapter is limited to the mechanical properties (tensile strength, Young's modulus, impact strength, and elongation at break) of semicrystalline/amorphous blends, and the attributing factors (e.g., morphology, blend-composition, interfacial interactions, crystallinity, dimensions of filler, etc.), for regulating the desired properties needed for applications. In the following section, classifications of polymers followed by multi-component polymer systems are discussed. Furthermore, challenges associated with the field of multi-component polymer systems will also be highlighted.

1.2 CLASSIFICATION OF POLYMERS

Polymers can be classified into thermoplastics, thermosets, and elastomers. Thermoplastics become soft and flexible upon heating and they solidify upon cooling. The transition of solid to liquid and vice versa is reversible by suitable heat treatment. Hence, thermoplastics can be easily recycled. In general, most of the thermoplastics are relatively linear polymer chains of high molecular weights, e.g., PE, poly(vinyl chloride) (PVC), Nylon 6, etc.

Thermosetting polymers become permanently hard when heat is applied on the materials and they do not soften upon subsequent heating. This is because, during the initial heat treatment, covalent cross-links are formed between adjacent molecular chains of the monomers; these bonds anchor the chains together to resist the vibrational and rotational chain motions at high temperature. Heating thermosets at excessively high temperatures will cause severance of these cross-link bonds and polymer degradation. They are generally harder and stronger than thermoplastics and have better dimensional stability. However, it is always very challenging to recycle thermosets. Most cross-linked network polymers come under thermosets, e.g., phenolic resin, urea/melamine resin, epoxies, and unsaturated polyesters.

Elastomers are flexible polymers with light cross-linking in the temperature range between its glass transition temperature (T_g) and its liquefaction (melting) temperature (T_m). Elastomeric properties appear when the backbone bonds can readily undergo torsional motions to permit uncoiling of the chains when the elastomers are stretched. Cross-links between the polymer chains prevent the chains from slipping past each other and thus prevent the elastomers from becoming permanently elongated when held under tension. They do not exhibit melting temperatures. Examples are NR, epoxidized natural rubber (ENR), styrene-butadiene rubber (SBR), chloroprene rubber (CR), EPDM, and nitrile-butadiene rubber (NBR).

On the other hand, polymers can be classified also into amorphous and semicrystalline, according to the alignment of the macromolecular chains in the polymer matrix. Semicrystalline polymers are those in which there are domains of both crystalline and amorphous structures. Figure 1.1(a) shows semicrystalline polymer with crystallites formed by aligned chain segments of long polymer chains, with other segments in the disordered matrix. The degree of crystallinity, molecular weight distribution, and the morphology of the semicrystalline polymers have profound effects on their physical and mechanical properties. Semicrystalline polymers exhibit apparent melting temperatures (T_m) at which the ordered regions become disordered. In general, they possess low viscosity above their melting point,[41] e.g., poly(ethylene oxide) (PEO), poly (ethylene terephthalate) (PET), high density poly(ethylene) (HDPE), (PCL), poly(lactic acid) (PLA), and PP.

Amorphous polymers have no three-dimensionally ordered repeating structure (c.f. Fig. 1.1(b)). Due to the lack of organization, no crystallinity is observed in amorphous polymers which leads to their temperature-dependent mechanical strengths. Amorphous polymers do not exhibit T_m. In general, amorphous polymers soften gradually upon heating. Below T_g, the amorphous polymer chains become immobilized, rigid and in glassy state, e.g., NR and PC.

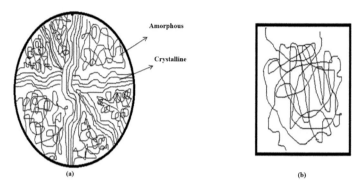

FIGURE 1.1 Alignment of macromolecular chains of (a) semicrystalline polymer and (b) amorphous polymer.

1.3 CLASSIFICATION OF MULTI-COMPONENT POLYMERIC SYSTEMS

Figure 1.2 describes the general classification used in this chapter for multi-component polymeric systems. Here, the classification includes binary polymer blends, polymer composites (one polymer with either organic or inorganic filler), and polymer blend composites.[42-58]

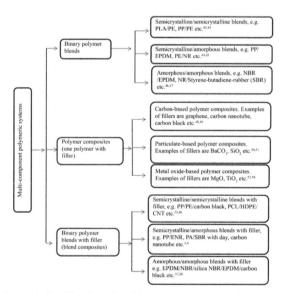

FIGURE 1.2 General classification of multi-component polymeric systems in this chapter.

1.4 MECHANICAL PROPERTIES

In this chapter, the mechanical properties (tensile strength, Young's modulus, elongation at break, and impact strength) of polymer blends, polymer composites, and polymer blend composites are discussed. For polymer blends, the addition of one polymer to another polymer can regulate the mechanical properties.[59,60] The main attributing factors in tuning the mechanical properties are polymer blend compositions, molecular weight of polymers, developed morphologies, interfacial adhesion of the components, and processing method. In the case of polymer composites, the additive filler plays a major role to switch the tensile strength, modulus, and elongation at break of the polymers. The compatibility of filler with the polymer matrix, dispersion and orientation of the filler in the polymer matrix are responsible in controlling the mechanical properties. Thereby, if the systems exhibit homogeneous dispersion of filler, and strong interaction between filler and the polymer matrix, improvement in mechanical properties of systems can be expected.[61] In the case of polymer blend composites, in addition to blend composition, the mechanical properties also depend upon the localization of filler in the blends, morphology, and processing method. Moreover, the compatibility between filler and the polymers plays a major role to adjust the properties of the systems.[62]

1.5 MECHANICAL PROPERTIES OF SEMICRYSTALLINE/ AMORPHOUS BINARY BLENDS

Improvement in mechanical properties of different binary semicrystalline/amorphous polymer blend systems is studied extensively.[63–72] In general, the variation in blend composition, chemical pre-treatment of parent components as well as processing strategy may enhance the mechanical properties of the systems. Mixing of polyolefins with amorphous polymers such as PC, natural rubber, and synthetic rubber could be a useful method to enhance the mechanical properties of polyolefins.[73–75] However, owing to the fact that lack of miscibility between the parent components, these blends will not display desired mechanical properties without compatibilization. Therefore, active surface modifiers are needed in most cases in order to enhance the compatibility between the constituents for the blends.

Extensive studies were carried out over decades for improvement of the toughness of PP by blending PP with amorphous polymer. However, in most cases, stiffness of PP is sacrificed when it is blended with another amorphous polymer. Hence, optimization of the toughness and stiffness for PP/amorphous blends with or without compatibilizing agent becomes the main focus of interest in most of the studies.[76–79]

Based on literature survey, the following factors may influence the mechanical properties of semicrystalline polymers when it is blended with various elastomers.[80–82]

1. The modulus of the elastomer is usually much less than that of the semicrystalline polymer; hence, lower tensile strength of the blends is normally recorded.
2. The uniform distribution of elastomer domains in the semicrystalline polymer may lead to improved impact strength of the blends.
3. A certain degree of interfacial adhesion attained between the elastomer particle and the semicrystalline polymer is also a reason for enhancing the impact strength.

Table 1.2 and Figure 1.3 (a)–(c) depict the mechanical properties of PP blended with various amorphous polymers which are either natural or synthesized elastomers. It can be seen that tensile strength and Young's modulus of the blends decrease with increasing rubber content. As the rubber content increases, coalescence of the rubber domains and the poor PP and rubber interface adhesion account for the observed decrease in tensile strength and Young's modulus of the polymer blends. This is a common drawback not only in the case of PP but also for other kinds of semicrystalline/elastomer blends.[83–85] Besides, the impact strength and elongation at break increase with the addition of rubber in PP matrix. The dynamically vulcanized PP/EPDM system with dimethylol phenolic resin as cross-linking agent shows maximum impact strength at 60 wt% of PP as compared to other blends due to the formation of co-continuous morphologies and the uniform distribution of EPDM domains in the PP matrix. Besides, the domain size of EPDM reduces because the cross-linking of EPDM in PP leads to homogeneous dispersion, thus causing consistent improvement in mechanical properties.[74]

Generally, the elongation at break[76] increases when the rubber content in PP increases because of the elasticity nature of rubber and the formation

A Review on Mechanical Properties of Semicrystalline

TABLE 1.2 Mechanical Properties of PP with Various Amorphous Polymers.

Blends and Compositions	Processing Methods and Conditions	Cross-linking Agent	Compatibilizer Used	Tensile Strength (MPa)	Elongation at Break (%)	Young's Modulus (MPa)	Impact Strength (J/m)	Ref.
NR				11.5	750			[104]
EPDM[a]				3.4	100			
NBR								
ENR-30								
PP/EPDM[b] (100/0)	Melt mixing by internal mixer 190 °C/80 rpm			33.2	46		25	
PP/EPDM (90/10)				23.5	52		39	
PP/EPDM (80/20)				20.4	66		57	[76]
PP/EPDM (70/30)				16.2	122		113	
PP/EPDM (60/40)				13.0	285		190	
PP/EPDM (90/10)	Melt mixing by internal mixer 190 °C/80 rpm	Dimethylol		24.6	56		58	
PP/EPDM (80/20)		Dimethylol		20.7	76		86	[76]
PP/EPDM (70/30)		Phenolic resin		17.8	140		935	
PP/EPDM (60/40)				13.9	350		1150	
PP/NBR[c] (80/20)	Melt mixing by internal mixer 185 °C/80 rpm			19.0	225			[77]
PP/NBR (60/40)				13.0	275			
PP/NBR (50/50)				10.7	319			
PP/NBR (40/60)				9.0	394			
		Sulfur		8.0	300			
PP/ENR- 30[d] (25/75)	Melt mixing by internal mixer 180 °C/60 rpm	Dicumyl peroxide Sulfur + dicumyl peroxide	Phenolic modified polypropylene (Ph-PP)	6.3 9.2	250 320			[78]

TABLE 1.2 (Continued)

Blends and Compositions	Processing Methods and Conditions	Cross-linking Agent	Compatibilizer Used	Tensile Strength (MPa)	Elongation at Break (%)	Young's Modulus (MPa)	Impact Strength (J/m)	Ref.
(PP/NR)100/0				37.8	12	1166		
(PP/NR) 80/20	Melt mixing by			21.3	30	975		[79]
(PP/NR) 70/30	internal mixer			17.8	42	733		
(PP/NR) 60/40	190 °C/50 rpm			12.5	55	610		
(PP/NR) 50/50				11.5	61	525		
(PP/NR) 40/60				6.0	101	345		
(PP/SBS) 95/5							40	
(PP/SBS) 90/10							47	
(PP/SBS) 85/15							79	[75]
(PP/SBS) 80/20							101	
(PP/SBS) 70/30							252	
(PP/SEBS) 95/5							38	
(PP/SEBS) 90/10							49	
(PP/SEBS) 85/15							72	
(PP/SEBS) 80/20	Single screw						107	
(PP/SEBS) 70/30	extruder						–	
(PP/EOC) 95/5							52	
(PP/EOC) 90/10	Twin screw						96	[107]
(PP/EOC) 85/15	extruder						267	
(PP/EOC) 80/20							363	
(PP/EOC) 70/30							155	

[a]Ethylene/propylene weight ratio = 73/27, diene content (mmol/100 g) = 58.9. [b]EPDM with ethylidene 2-norbonene (ENB) E/P weight ratio = 74/26, ENB = 5 wt %. [c]ACN content (%) 34 ± 2. [d]ENR-30 with a level of epoxide groups at approximately 30 mol% epoxide. [e]Ethylene-octene copolymer.

of cross-links after curing. PP/ENR-30 system with different curing agents like sulfur, dicumyl peroxide (DCP), and mixed system (sulfur combined with DCP) were examined with respect to chemical interactions (bonding between carbon and sulfur) on mechanical properties. Among the three curative agents, mixed system shows enhanced mechanical properties owing to the formation of C–C, S–S, and C–S bonds during the mixed curing process (sulfur + DCP).

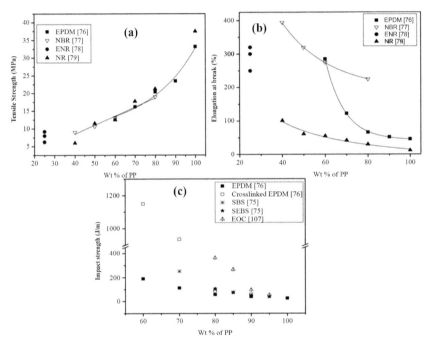

FIGURE 1.3 Mechanical properties of PP with various rubbers: (a) tensile strength, (b) elongation at break, and (c) impact strength.

The mechanical properties of high density poly(ethylene) (HDPE) with various rubbers[83,84,86] are summarized in Table 1.3. Tensile strength and Young's modulus of the HDPE-based blends decrease with increasing rubber content and the elongation at break increases with the addition of rubber as expected. Besides, the blends (HDPE/SBR, HDPE/NBR) show acceptable tensile strength (15 MPa) at 75 wt% of HDPE as compared to pure HDPE (32 MPa). This is because of the excess amount of total thermoplastic content. The effect

TABLE 1.3 Mechanical Properties of HDPE with Various Amorphous Polymers.

Blends & Compositions	Processing Methods & Conditions	Cross-linking Agent	Compatibilizer Used	Tensile Strength (MPa)	Elongation at Break (%)	Young's Modulus (MPa)	Ref.
HDPE				32	150		[105]
HDPE/ENR-30[a] (50/50)	Melt mixing by internal mixer 180 °C/60 rpm	Sulfur and DCP	Phenolic modified HDPE compatibilizer (PhHRJ-HDPE)	8.4	211		[83]
HDPE/ENR-30 (40/60)				7.8	269		
HDPE/ENR-30 (25/75)				7.3	331		
HDPE/SBR[b] (75/25)	Melt mixing by internal mixer 130 °C/30 rpm			16.3	150	13.8	
HDPE/SBR (50/50)				11.3	200	8.3	
HDPE/SBR (25/75)				5.9	225	3.7	
HDPE/NBR[c] (75/25)				13.0	400	9.9	
HDPE/NBR (50/50)				13.8	485	9.0	[84]
HDPE/NBR (25/75)				7.2	255	4.4	
HDPE/NR (40/60)	Melt mixing by internal mixer 180 °C/60 rpm		Hydroxyl methyl phenolic resin	9.0	300		[86]
			Dimethyl phenolic resin	8.9	100		

[a]Level of epoxide groups at 30%.

[b]Styrene butadiene rubber (butadiene styrene copolymer with a styrene content of 23.5%).

[c]Nitrile rubber (butadiene acrylonitrile copolymer with 32% acrylonitrile).

of compatibility was analyzed by hydroxyl methyl phenolic resin (HRJ) and dimethyl phenolic resin (SP) in the HDPE/NR systems. In which, HRJ modified system exhibits higher elongation at break than SP modified system. The main expected reason for the enhancement of compatibility via the π–π interactions among NR domains in the HDPE and HRJ leads to high elongation at break. The results presented in Figure 1.4 reflect the variations in the mechanical properties of the different HDPE/amorphous blends.

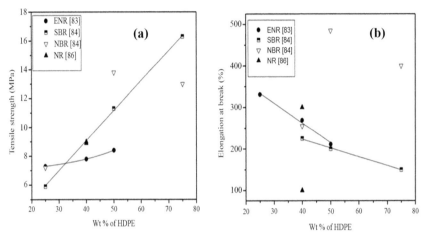

FIGURE 1.4 Mechanical properties of HDPE/amorphous blends: (a) tensile strength and (b) elongation at break.

Nylon is a poly amide; it consists of amide linkages formed by the acid and amino groups. Nylon has amazing industrial importance owing to its high mechanical strength, electrical insulation, high mechanical damping, and excellent wear resistance. It can be used as slide bearings, gear wheels, synthetic fibers, insulators, etc. Nylon is the predominant material for life safety ropes owing to its high durability and cost effectiveness. However, a few drawbacks are noticed while using nylon for certain applications. Nylon can be damaged by various chemicals including acids and it losses approximately 15% of its strength when it gets wet but regains its strength when dried. In order to avoid this problem, nylon has to be blended with various other polymers comprising semicrystalline and amorphous polymers.

In general, nylons are categorized into six groups according to the number of carbon atoms present in the monomeric unit and methylene groups.

Group-1	Diamines with even number of methylene groups and dibasic acid with even number of methylene groups; e.g., Nylon 66; Nylon 6,10; Nylon 4,8
Group-2	Diamine with odd number of methylene groups and dibasic acid with even number of methylene groups; e.g., Nylon 7,6; Nylon5,6; Nylon 7,8
Group-3	Diamine with even number of methylene groups and dibasic acid with odd number of methylene group; e.g., Nylon 4,9; Nylon 6,9; Nylon 8,9
Group-4	Diamine with odd number of methylene groups and dibasic acid with odd number of methylene group; e.g., Nylon 5,7, Nylon 7,9, Nylon 5,9
Group-5	Amino acids with odd number of methylene groups and even numbers of carbon atom in the repeat; e.g., Nylon 4; Nylon 6; Nylon 10
Group-6	Amino acids with even number of methylene groups and odd number of carbon atom in the repeat; e.g., Nylon 7; Nylon 11

In general, most of the nylon/amorphous polymer blends are immiscible; many interesting studies were carried out on mechanical properties of nylon by blending with various amorphous polymers with respect to their morphologies.[87–94] To meet desirable properties and commodity applications with these blends, compatibilizer should be added during the blending process owing to lack of interfacial adhesion and compatibility. However, nylon can exhibit excellent toughness with amorphous polymers where there could be a strong chemical interaction between amorphous polymer and nylon. For example, toughening of nylon with epoxidized ethylene propylene diene rubber (eEPDM) was observed by Wang et al.[108] The main attributing factor for enhancing the toughness of nylon is that the grafted epoxy groups of EPDM can react with the amino end group of nylon to form a grafted copolymer which could act as a compatibilizer between nylon and eEPDM. The effect of compatibilizer on the mechanical properties of nylon/EPDM blends was reported by Huang et al.[95] They used different compatibilizers like maleic anhydride grafted EPR (MAH-g-EPR), maleic anhydride grafted EPDM (MAH-g-EPDM), and carboxyl terminal nitrile rubber. Among these compatibilizers, MAH-g-EPR shows high tensile strength (13.3 MPa) as compared to MAH-g-EPDM (10.1 MPa), and carboxyl terminal nitrile rubber (4.2 MPa). This is due to the polar part of the MAH-g-EPR which could easily interact with nylon through hydrogen bonding (chemical interaction) whereas the non-polar EPR of MAH-g-EPR interacts with EPDM through π–π interactions.

The mechanical properties of Nylon 6 with various amorphous polymers with and without compatibilizers are listed in Table 1.4.[85,96,97] From Table 1.4, it is clear that the tensile strength and Young's modulus decrease with the addition of amorphous polymer. Besides, the impact strength increases with the addition of amorphous polymer. Herein, Nylon 6 with SEBS grafted ε-caprolactam blocked allyl(3-isocyanate-4-tolyl)carbamate (SEBS-g-BTAI) shows maximum impact strength. The formation of new copolymer at the interphase of the Nylon 6 and SEBS is the main reason for the improvement in properties. The main reason for supporting this explanation is that the core shell morphologies due to the formation of copolymer at the interface are observed by FE-SEM images in Figure 1.5. The Nylon 6/ABS blends exhibit high Young's modulus values compared to other blends. This may be because of high individual strength of ABS compared to other elastomers. The results obtained are also depicted in Figure 1.6.

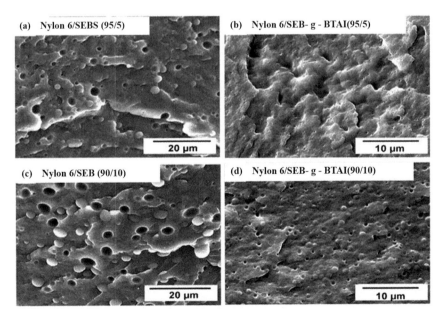

FIGURE 1.5 FESEM micrographs of the Nylon 6/SEBS blends with and without compatibilizer.[85] (Reprinted from Yin, L.; Yin, J.; Shi, D.; Luan, S. Effects of SEBS-g-BTAI on the Morphology, Structure and Mechanical Properties of PA6/SEBS Blends. Eur. Polym. J. 2009, 45, 1554–1560. © 2009 with permission from Elsevier.)

TABLE 1.4 Mechanical Properties of Nylon 6 with Various Amorphous Polymers.

Blends and Compositions	Processing Methods and Conditions	Compatibilizer Used	Tensile Strength (MPa)	Elongation at Break (%)	Young's Modulus (MPa)	Impact Strength (J/m)	Ref.
Nylon 6			59.4		706	50	
Nylon 6/SEBS[a] (95/5)	Melt mixing by internal mixer		49.6		661	52	[85]
Nylon 6/SEBS (90/10)			43.2		659	56	
Nylon 6/SEBS (87.5/12.5)	235 °C/70 rpm		40.5		645	63	
Nylon 6/SEBS (85/15)			39.5		633	77	
Nylon 6/SEBS (80/20)		ε-caprolactam blocked allyl (3-isocyanate-4-tolyl) carbamate (BTAI)	26.8		477	81	
Nylon 6/SEBS (95/5)			52.4		682	161	
Nylon 6/SEBS (90/10)			47.2		691	303	
Nylon 6/SEBS (87.5/12.5)			46.3		663	362	[85]
Nylon 6/SEBS (85/15)			43.5		644	1054	
Nylon 6/SEBS (80/20)			41.0		635	1082	
Nylon 6	Twin-screw extruder		74.3	45			
Nylon 6/ENR[b] (70/30)			28.3	60			[96]
Nylon 6/NR (70/30)			45.2	51			
Nylon 6/ABS[c] (0/100)					2250	335	
Nylon 6/ABS (15/75)					2270	380	
Nylon 6/ABS (25/65)	Twin-screw extruder	Styrene-acrylonitrile glycidyl methacrylate (SAGMA)			2290	415	[97]
Nylon 6/ABS (35/55)					2355	455	
Nylon 6/ABS (45/45)					2380	580	
Nylon 6/ABS (55/35)					2390	715	

A Review on Mechanical Properties of Semicrystalline

TABLE 1.4 *(Continued)*

Blends and Compositions	Processing Methods and Conditions	Compatibilizer Used	Tensile Strength (MPa)	Elongation at Break (%)	Young's Modulus (MPa)	Impact Strength (J/m)	Ref.
Nylon 6/ABS (50/50)				103	1470	106	
Nylon 6/ABS (47.5/47.5)				87	1590	947	
Nylon 6/ABS[d] (55/40)	Single screw extruder	Imidized acrylic (A)		60	1690	938	[106]
Nylon 6/ABS (60/35)				80	1830	913	
Nylon 6/ABS (65/30)	240 °C/40 rpm			137	1900	867	
Nylon 6/ABS (70/25)				125	1920	646	
Nylon 6/ABS (75/20)				83	2140	252	

[a]SEBS: styrene-b-(ethylene-co-1-butene)-b-styrene. Styrene content was 29 wt%.

[b]Epoxidized natural rubber containing 50 mol% of epoxidation.

[c]ABS Cycolac EX 10 U.

[d]ABS Acrylo nitrile content 24%, rubber content 45%.

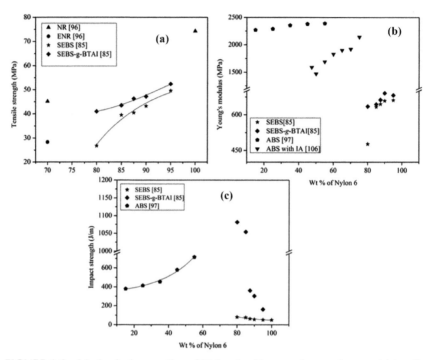

FIGURE 1.6 Mechanical properties of Nylon 6 with amorphous polymers: (a) tensile strength, (b) impact strength, and (c) Young's modulus.

In general, PC is an amorphous and thermoplastic polymer and it exhibits exceptional properties such as impact strength, toughness, and ductility. But the problem with PC is its high cost. Besides, it exhibits high shear viscosity and low melt strength and is thus difficult to process. Hence, in order to reduce the cost of PC and to avoid difficulty in processing, PC is to be mixed with other cheaper polymeric components. However, it is important to add the blends of PC with various semicrystalline polymers where chemical interactions like hydrogen bonding and "dipole dipole" interactions play a vital role in order to improve the mechanical, gas barrier, and also thermal properties. Compatibilizer should often be added in order to enhance the compatibility and to meet desirable properties. In most of the investigations, glycidyl methacrylate (GMA) is used as a compatibilizer for improving the compatibility of PC with various functionalized poly olefins. The epoxy groups of GMA can react with the corresponding reactive groups, such as –OH, –COOH, –NH$_2$, etc. A series of investigations were carried out with respect to the effect of different compatibilizers on the mechanical properties of PC based blends.[98-100] Zhihui et al.[101] reported on the effect of compatibilizers on the mechanical properties of PP/PC blends. They reported that the addition of compatibilizer can significantly regulate the blends morphologies and crystallinity of pure PP in the PC.

Comparative investigations on the mechanical properties of PC with various semicrystalline polymers with and without compatibilizers are listed in Table 1.5.[101-103] From Table 1.5, it can be noticed that the maximum tensile strength values are achieved in the range of 40–60 wt% of PC with the addition of various semicrystalline polymers. The main attributing factors for this improvement are the decrease in the rate of crystallinity of pure semicrystalline polymers by blending with PC and the shrinkage of crystalline domains in the amorphous PC promotes the high stress transfer at the interface. For example, the PP-grafted-GMA (PP-g-GMA) compatibilized PC/PP system exhibits high mechanical properties than unmodified system. This is mainly because of PP-g-GMA, which acts as an interfacial agent and enhances interfacial adhesion among PC domains in the PP phase. It also reduces the interfacial tension and domain size of PC. However, there is no clear trend in elongation at break with the addition of PC. But, certain blend ratios (40–60 wt% of PC) exhibit almost constant elongation at break. Due to the formation of co-continuous morphologies, elongation is almost the same when stress is applied. Besides, we also illustrate the tensile strength and elongation at break in Figure 1.7.

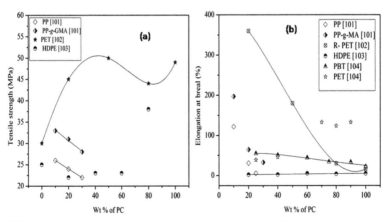

FIGURE 1.7 Mechanical properties of PC with semicrystalline polymers: (a) tensile strength and (b) elongation at break.

TABLE 1.5 Mechanical Properties of Various Semicrystalline Polymers with PC.

Blends and Compositions	Processing methods and Conditions	Compatibilizer Used	Tensile Strength (MPa)	Elongation at Break (%)	Ref.
PP/PC (90/10)	Twin-screw extruder		26	122.0	
PP/PC (80/20)	250 °C/100 rpm		24	31.0	
PP/PC (70/30)			22	5.7	[101]
PP/PC (60/30)		PP-g-GMA (5%)	28	32.6	
PP/PC (80/10)	Twin-screw extruder	PP-g-GMA (10%)	33	197	
PP/PC (70/20)	250 °C/100 rpm	PP-g-GMA (10%)	31	64.5	
PP/PC (60/30)		PP-g-GMA (10%)	36	49.7	
PP/PC (50/30)		PP-g-GMA (20%)	42	69.4	
PC			49	16	
PETª/PC (20/80)			44	30	
PET/PC (50/50)	Twin-screw extruder	catalyst (Tin-ethyl-hexanoate)	50	180	[102]
PET/PC (80/20)	100–110 rpm		45	360	
PET			30	60	
HDPE/PC (100/0)	Melt mixing by		25	7.8	
HDPE/PC (80/20)	Internal mixer	—	22	2.2	[103]
HDPE/PC (60/40)	250 °C/50 rpm		23	2.5	
HDPE/PC (40/60)			23	5.3	
HDPE/PC (20/80)			38	4.1	
HDPE/PC (0/100)			60	5.0	

TABLE 1.5 *(Continued)*

Blends and Compositions	Processing methods and Conditions	Compatibilizer Used	Tensile Strength (MPa)	Elongation at Break (%)	Ref.
PBT[b]/ PC (100/0)	Single-extruder		56	31.0	
PBT/ PC (75/25)	250 °C		59	55.0	
PBT/ PC (60/40)			61	51.0	[104]
PBT/ PC (40/60)			65	44.0	
PBT/ PC (25/75)			66	34.0	
PBT/ PC (10/90)			65	33.0	
PBT/ PC (0/100)			57	22.0	
PET/PC (100/0)	Single-extruder		55	28.0	
PET/PC (75/25)	275 °C		57	39.0	
PET/PC (60/40)			60	46.0	[104]
PET/PC (30/70)			65	133.0	
PET/PC (20/80)			64	124.0	
PET/PC (10/90)			57	133.0	
PET/PC (0/100)			57	22.0	

[a]Recycled poly(ethylene terephthalate) PET.

[b]Poly(butylene terephthalate) PBT.

1.6 CONCLUSION AND FUTURE PERSPECTIVES

In this chapter, we have made an attempt to describe various polymers being used for making multi-component polymeric systems. Along with the fundamental classification of polymers and multi-component polymeric systems we have also covered the mechanical properties of semi-crystalline/amorphous polymer blends (PP, HDPE, nylon, PET, PBT, etc, with different amorphous polymers) including tensile strength, Young's modulus, impact strength, and elongation at break. The most interesting and important part of this chapter is the correlation of the mechanical properties with the morphology, chemical and physical interactions, and the polymer blend ratio.

We have found that the tensile strength and Young's modulus of all the semicrystalline/elastomer blends decrease with the addition of elastomer and was found to be due to the low modulus values of elastomers. The elongation at break increases with the addition of elastomer owing to the fact that elastomer tends to show high elongation as compared to the semi-crystalline polymers which was mainly contributed by the elastic behavior of the amorphous domain. Besides, the presence of compatibilizer along with the curing agent tends to decrease the elongation at break owing to the formation of interconnections through covalent bonds between macro molecular chains resulting in the restriction of chain mobilization. Impact strength values were reported to be increased with elastomer loading when there is a homogeneous distribution of elastomer components inside the semicrystalline polymer. Moreover, the elastomer components may absorb high mechanical energy leading to high stress transfer. The high damping nature of elastomers is another reason for the increase in impact strength. The reported works show that the optimum impact values of the blends were observed when we have a polymer blend with co-continuous morphology.

The role of miscibility, microstructure, and phase behavior of polymer blends on the mechanical properties still needs to be explored. The use of computational modeling to estimate the behavior of polymer blends should be undertaken. Though a few computer software methods like QSPR (Quantitative Structure Property Relationship) and PRISM (Statistical Mechanical Polymer Reference Interaction Site Model) have been developed to predict the miscibility and phase behavior of binary polymer blends, we have to go a long way to make good progress in this direction.

KEYWORDS

- **mechanical properties**
- **polymer blends**
- **multi-component polymeric systems**
- **semicrystalline/amorphous binary systems**
- **thermoplastics**
- **thermosets**
- **elastomers**

REFERENCES

1. Loyens, W.; Groeninckx, G. Ultimate Mechanical Properties of Rubber Toughened Semicrystalline Poly(Ethylene Terephthalate) at Room Temperature. *Polymer*. **2002**, *43*, 5679–5691.
2. Sharma, R.; Joshi, H.; Jain, P. Elastomer Toughened Poly(Butylene Terephthalate) Nanocomposites Morphology and Impact Strength. *Arch. Appl. Sci. Res*. **2012**, *4*, 1833–1838.
3. Gonzalez, I.; Eguiazabal, J. I.; Nazabal, J. Rubber-Toughened Polyamide 6/Clay Nanocomposites. *Compos. Sci. Technol*. **2006**, *66*, 1833–1843.
4. Gonzalez, I.; Eguiazabal, J. I; Nazabal, J. Attaining High Electrical Conductivity and Toughness in Pa6 by Combined Addition of MWCNT and Rubber. *Compos. Part A-Appl*. **2012**, *43*, 1482–1489.
5. Liu, X.; Wu, Q. PP/Clay Nano Composites Prepared by Grafting-Melt Intercalation. *Polymer*. **2001**, *42*, 10013–10019.
6. Ahmadi, S. J.; Sell, C.; Huang, Y.; Ren, N.; Mohaddespour, A.; Hiver, J. M. Mechanical Properties of NBR/Clay Nanocomposites by Using a Novel Testing System. *Compos. Sci. Technol*. **2009**, *69*, 2566–2572.
7. Keskin, R.; Adanur, S. Improving Toughness of Polypropylene with Thermoplastic Elastomers in Injection Molding. *Polym. Plast. Technol. Eng.* **2011**, *50*, 20–28.
8. Abreu, F. O. M. S.; Forte, M. M. C.; Liberman, S. A. SBS and SEBS Block Copolymers as Impact Modifiers for Polypropylene Compounds. *J. Appl. Polym. Sci*. **2005**, *95*, 254–26.
9. Zuiderduin, W. C. J.; Westzaan, C.; Huetink, J.; Gaymans, R. J. Toughening of Polypropylene with Calcium Carbonate Particles. *Polymer*. **2003**, *44*, 261–275.
10. Mahfuz, H.; Adnan, A.; Rangari, V. K.; Jeelani, S.; Jang, B. Z. Carbon Nanoparticles/Whiskers Reinforced Composites and their Tensile Response. *Compos Part A-Appl. Sci. Eng*. **2004**, *35*, 519–527.

11. Koning, C.; Duin, M. V.; Pagnoulle, C.; Jerome B. R. Strategies for Compatibilization of Polymer Blends. *Prog. Polym. Sci.* **1998**, *23*, 707–757.
12. Xing, P.; Dong, L.; An, Y.; Feng, Z. Miscibility and Crystallization of Poly(B-Hydroxybutyrate) and Poly(P-Vinylphenol) Blends. *Macromolecules.* **1997**, *30*, 2726–2733.
13. Urakawa, O.; Fuse, Y.; H. Hori, H.; Tran-Cong, Q.; Yano, O. A Dielectric Study on the Local Dynamics of Miscible Polymer Blends: Poly(2-Chlorostyrene)/Poly(Vinyl Methyl Ether). *Polymer.* **2001**, *42*, 765–773.
14. Silvestre, C.; Cimmino, S.; Di Pace, E. *Crystallization Polymer Blends; Polymeric Materials Encyclopedia;* CRC Press: New York, 1996; Vol. 2.
15. Nakason, C.; Saiwari, S.; Kaesaman, A. Rheological Properties of Maleated Natural Rubber/Polypropylene Blends with Phenolic Modified Polypropylene and Polypropylene-G-Maleic Anhydride Compatibilizers. *Polym. Test.* **2006**, *25*, 413–423.
16. Tomar, N.; Maiti, S. N. Mechanical Properties of PBT/ABAS Blends. *J. Appl. Polym. Sci.* **2007**, *104*, 1807–1817.
17. Wang, C.; Su, J. X.; Li, J.; Yang, H.; Zhang, Q.; Du, R. N.; Fu, Q. Phase Morphology and Toughening Mechanism of Polyamide 6/EPDM-G-MA Blends Obtained Via Dynamic Packing Injection Molding. *Polymer.* **2006**, *47*, 3197–3206.
18. Olabisi, O.; Robeson, L. M.; Shaw, T. *Polymer-Polymer Miscibility;* Academic Press: New York, 1979.
19. Utracki, L. A. Introduction to Polymer Blends. In *Polymer Blends Handbook;* Kluwer Academic Press: The Netherlands, 2002; Vol. 1.
20. Utracki, L. A. *Two Phase Polymer Systems;* Hanser Gardner: Munich, 1991.
21. Dumoulin, M. M. Polymer Blends Forming. In *Polymer Blends Handbook;* Kluwer Academic Press: The Netherlands, 2002; Vol. 2.
22. Utracki, L. A. *Polymer Blends Handbook;* L. A., Utracki, Ed., Kluwer Academic Publishers: Dordrecht, The Netherlands, 2002.
23. Yin, B.; Zhao, Y.; Pan, M.; Yang, M. Morphology and Thermal Properties of a PC/PE Blend with Reactive Compatibilization. *Polym. Adv. Technol.* **2007**, *18*, 439–445.
24. Nakason, C.; Saiwari, S.; Kaesaman, A. Rheological Properties of Maleated Natural Rubber/Polypropylene Blends with Phenolic Modified Polypropylene and Polypropylene-G-Maleic Anhydride Compatibilizers. *Polym. Test.* **2006**, *25*, 413–423.
25. Chatterjee, K.; Naskar, K. Development of Thermoplastic Elastomers Based on Maleated Ethylene Propylene Rubber (M-EPM) and Polypropylene (PP) by Dynamic Vulcanization. *Express Polym. Lett.* 2007, *1*, 527–534.
26. Loyens, W.; Groeninckx, G.; Ultimate Mechanical Properties of Rubber Toughened Semicrystalline PET at Room Temperature. *Polymer.* **2002**, *43*, 5679–5691.
27. Eklind, H.; Schantz, S.; Maurer, F. H. J. Characterization of the Interphase in PPO/PMMA Blends Compatibilized by P(S-*g*-EO). *Macromolecules.* **1996**, *29*, 984–992.
28. Grigoraş, V. C.; Bărboiu, V. Characteristics of Compatible Binary Polymer Blends Deduced from DSC Thermograms. 1. A Study on Polyvinyl Alcohol - Polyvinyl Pyrrolidone Mixtures. *Rev. Roum. Chim.* 2008, *53*, 127–131.
29. Halary, J. L.; Jarray, J.; Fatnassi, M.; Larbi, F. B. C. Amorphous and Semicrystalline Blends of Poly(Vinylidene Fluoride) and Poly(Methyl Methacrylate):Characterization and Modeling of the Mechanical Behavior. *J. Eng. Mater. Technol.* **2011**, *134*, 0109101–109108.

30. Rim, P. B.; Runt, J. P. Melting Behavior of Crystalline/Compatible Polymer Blends: Poly(E-Caprolactone)/Poly(Styrene-Co-Acrylonitrile). *Macromolecules.* **1983,** *16,* 762–768.

31. Ismail, H.; Suryadiansyah. Thermoplastic Elastomers Based on Polypropylene/ Natural Rubber and Polypropylene/Recycle Rubber Blends. *Polym. Test.* 2002, *21,* 389–395.

32. Oh, J. S.; Isayev, A. I.; Rogunova, M. A. Continuous Ultrasonic Process for in Situ Compatibilization of Polypropylene/Natural Rubber Blends. *Polymer.* **2003,** *44,* 2337–2349.

33. Bhattacharya, S. N.; Kamal, M.; Gupta, R. *Polymeric Nanocomposites. Theory and Practice;* Carl Hanser Verlag: Munich, Germany, 2008.

34. Ray, S. S.; Okamoto, M. Polymer/Layered Silicate Nanocomposites: A Review from Preparation to Processing. *Prog. Polym. Sci.* **2003,** *28,* 1539–1641.

35. Aurilia, M.; Sorrentino, L.; Berardini, F.; Sawalha, S.; Iannace, S. Mechanical Properties of Nano/Micro Multilayered Thermoplastic Composites Based on PP Matrix. *J. Thermoplast. Compos.* **2011,** *25,* 835–849.

36. Yu, Z.; Ou, Y.; Qi, Z.; Hu, G. Toughening of Nylon 6 with a Maleated Core-Shell Impact Modifier. *J. Polym. Sci. Pol. Phys.* **1998,** *36,* 1987–1994.

37. Gonzalez, I.; Eguiazabal, J. I.; Nazabal, J. Rubber-Touhened Polyamide 6/Clay Nanocomposites. *Compos. Sci. Technol.* **2006,** *66,* 1833–1843.

38. Lim, J. W.; Hassan, A.; Rahmat, A. R.; Wahit. M. U. Rubber-Toughened Polypropylene Nanocomposite: Effect of Polyethylene Octane Copolymer on Mechanical Properties and Phase Morphology. *J. Appl. Polym. Sci.* **2006,** *99,* 3441–3450.

39. Wang, Y.; Zhang, Q.; Fu, Q.; Compatibilization of Immiscible Poly(Propylene)/Polystyrene Blends Using Clay. *Macromol. Rapid Commun.* **2003,** *24,* 231–235.

40. Ray, S. S.; Pouliot, S.; Bousmina, M.; Utracki, L. A. Role of Organically Modified Layered Silicate as an Active Interfacial Modifier in Immiscible Polystyrene/Polypropylene Blends. *Polymer.* **2004,** *45,* 8403–8413.

41. Silvestre, C.; Cimmino, S.; Di pace, E. *Polymeric Materials Encyclopedia;* CRC Press: Boca Raton, New York, 1996.

42. Anderson, K. S.; Hillmyer, M. A. The Influence of Block Copolymer Microstructure on the Toughness of Compatibilized Polylactide/Polyethylene Blends. *Polymer.* **2004,** *45,* 8809–8823.

43. Loa, C. T.; Seifert, S.; Thiyagarajan, P.; Narasimhan, B. Phase Behavior of Semicrystalline Polymer Blends. *Polymer.* **2004,** *45,* 3671–3679.

44. Frounchi, M.; Dadbin, S.; Salehpour, Z.; Noferesti, M. Gas Barrier Properties of PP/ EPDM Blend Nanocomposites. *J. Membrane. Sci.* **2006,** *282,* 142–148.

45. Haque, M. A.; Ahmad, M. U.; Khan, M. A.; Raihan, S. M. A.; Dafader, N. C. Studies on the Physicochemical Properties of Natural Rubber/Polyethylene Blends and the Impact of Radiation on their Properties. *Polym. Plast. Technol.* **2010,** *49,* 1010–1015.

46. Al-Gahtani, S. A. Mechanical Properties of Acrylonitrile Butadiene/Ethylene Propylene Diene Monomer Blends: Effects of Blend Ratio and Filler Addition. *J. Am. Sci.* **2011,** *7,* 804–809.

47. Goyanes, S.; Lopez, C. C.; Rubiolo, G. H.; Quasso F.; Marzocca, A.J. Thermal Properties in Cured Natural Rubber/Styrene Butadiene Rubber Blends. *Eur. Polym. J.* **2008,** *44,* 1525–1534.

48. Omar, A.; Hartomy, A.; Ghamdi, A.; Dishovsky, N.; Shtarkova, R.; Iliev, V.; Mutlay, I.; Tantawy, F. Dielectric and Microwave Properties of Natural Rubber Based Nanocomposites Containing Graphene. *Mater. Sci. Appl.* **2012,** *3,* 453–459.

49. Du, J.; Zhao, L.; Zeng, Y.; Zhang, L.; Li, F.; Liu, P.; Liu, C. Comparison of Electrical Properties between Multi-Walled Carbon Nanotube and Graphene Nanosheet/High Density Polyethylene Composites with a Segregated Network Structure. *Carbon.* **2011,** *49,* 1094–1100.

50. Fu, S. H.; Feng, X. Q.; Lauke, B.; Mai, Y. W. Effects of Particle Size, Particle/Matrix Interface Adhesion and Particle Loading on Mechanical Properties of Particulate-Polymer Composites. *Compos. Part B- Eng.* **2008,** *39,* 933–961.

51. Wang, X.; Tong, W.; Li, W.; Huang, H.; Yang, J.; Li, G. Preparation and Properties of Nanocomposite of Poly(Phenylene Sulfide)/Calcium Carbonate. *Polym. Bull.* **2006,** *57,* 953–962.

52. Cho, S.; Choi, W. Solid-Phase Photocatalytic Degradation of PVC-Tio$_2$ Polymer Composites. *J. Photochem. Photobio. A.* **2001,** *143,* 221–228.

53. Hamming, L. M.; Qiao, R.; Messersmith, P. B.; Brinson. L. C. Effects of Dispersion and Interfacial Modification on the Macroscale Properties of Tio$_2$ Polymer-Matrix Nanocomposites. *Compos. Sci. Technol.* **2009,** *69,* 1880–1886.

54. Kovtyukhova, N.; Ollivier, P. J.; Chizhik, S.; Dubravin, A.; Buzaneva, E.; Gorchinskiy, A.; Marchenko, A.; Smirnova. N. Self-Assembly of Ultrathin Composite Tio$_2$ / Polymer Films. *Thin Solid Films.* **1999,** *337,* 166–170.

55. Shen, L.; Wang, F. Q.; Yang, H.; Meng Q. R. The Combined Effects of Carbon Black Andcarbon Fiber on the Electrical Properties of Composites Based on Polyethylene or Polyethylene/Polypropylene Blend. *Polym. Test.* **2011,** *30,* 442–448.

56. Ruelle, B.; Peeterbroeck, S.; Bittencourta, C.; Gorrasic, G.; Patimod, G.; Michel Hecqa, M.; Snydersa, R.; Pasqualed, S.; Philippe Duboisa, P. Semi-Crystalline Polymer/Carbon Nanotube Nanocomposites: Effect of Nanotube Surface-Functionalization and Polymer Coating on Electrical and Thermal Properties. *React. Funct. Polym.* **2012,** *72,* 383–392.

57. Suzana S. J.; Jovanovic, V.; Markovic, G.; Konstantinovic, S.; Milena M. C. Nanocomposites Based on Silica-Reinforced Ethylene-Propylene-Diene-Monomer/Acrylonitrile-Butadiene Rubber Blends. *Compos. Part B-Eng.* **2011,** *42,* 1244–1250.

58. Sau, K. P.; Chaki, T. K.; Khastgir, D. Carbon Fibre Filled Conductive Composites Based on Nitrile Rubber (NBR), Ethylene Propylene Diene Rubber (EPDM) and their Blends. *Polymer.* **1998,** *39,* 6461–6471.

59. Robeson, L. M. *Polymer Blends: A Comprehensive Review;* Carl Hanser Verlag: Munich, Germany, 2007.

60. Awang, M.; Ismail, H.; Hazizan, M. A. Polypropylene-Based Blends Containing Waste Tire Dust: Effects of Trans-Polyoctylene Rubber (TOR) and Dynamic Vulcanization. *Polym. Test.* **2007,** *26,* 779–787.

61. Rattanasom, N.; Prasertsri, S. Mechanical Properties, Gas Permeability and Cut Growth Behaviour of Natural Rubber Vulcanizates: Influence of Clay Types and Clay/Carbon Black Ratios. *Polym. Test.* **2012,** *31,* 645–653.

62. Mirzadeh, A.; Lafleur, P. G.; Kamal, M. R.; Dubois, C. The Effect of Compatibilizer on the Co-Continuity and Nanoclay Dispersion Level of TPE Nanocomposites Based on PP/EPDM. *Polym. Eng. Sci.* **2010,** *50,* 2131–2142.

63. Thitithammawong, A.; Nakason, C.; Sahakaro, K.; Noordermeer, J. Effect of Different Types of Peroxides on Rheological, Mechanical, and Morphological Properties of Thermoplastic Vulcanizates Based on Natural Rubber/Polypropylene Blends. *Polym. Test.* **2007,** *26,* 537–546.
64. Mohamed, M. G.; Abd-EL-Messieh, S. L.; EL-Sabbagh, S.; Younan, A. F. Electrical and Mechanical Properties of Polyethylene-Rubber Blends. *J. Appl. Polym. Sci.* **1998,** *69,* 775–783.
65. Mahaprama, S.; Poompradub, S. Preparation of Natural Rubber (NR) Latex/Low Density Polyethylene (LDPE) Blown Film and its Properties. *Polym. Test.* **2011,** *30,* 716–725.
66. Zebarjada, S. M.; Lazzerib, A.; Bagheri, R.; Reihani, M.; Frounchi, S. M.; Zebarjad, S. M. Fracture Mechanism Under Dynamic Loading of Elastomer-Modified Polypropylene. *Mater Lett.* **2003,** *57,* 2733–2741.
67. Wang, W.; WU, Q.; Baojun qu. Mechanical Properties and Structural Characteristics of Dynamically Photocrosslinked PP/EPDM Blends. *Polym. Eng. Sci.* **2003,** *43,* 1798–1805.
68. Echevarrıa, G.; Eguiazabal, J. I.; Nazabal, J. Influence of Compatibilization on the Mechanical Behavior of Poly(Trimethylene Terephthalate)/Poly(Ethylene–Octene) Blends. *Eur. Polym. J.* **2007,** *43,* 1027–1037.
69. Loyens, W.; Groeninckx, G. Deformation Mechanisms in Rubber Toughened Semicrystalline Polyethylene Terephthalate. *Polymer.* **2003,** *44,* 4929–4941.
70. Jha, A.; Bhowmick, A. K. Mechanical and Dynamic Mechanical Thermal Properties of Heat- and Oil-Resistant Thermoplastic Elastomeric Blends of Poly(Butylene Terephthalate) and Acrylate Rubber. *J. Appl. Polym. Sci.* **2000,** *78,* 1001–1008.
71. Phinyocheep, P.; Saelao, J. Buzare, J. Y. Mechanical Properties, Morphology and Molecular Characteristics of Poly(Ethyleneterephthalate) Toughened by Natural Rubber. *Polymer.* **2007,** *48,* 5702–5712.
72. Jha, A.; Bhowmick, A. K. Thermoplastic Elastomeric Blends of Poly(Ethylene Terephthalate) and Acrylate Rubber: 1 Influence of Interaction on Thermal, Dynamic Mechanical and Tensile Properties. *Polymer.* **1997,** *38,* 4337–4344.
73. Li, Z. M.; Huang, C. G.; Yang, W.; Yang, M. B.; Huang, R. Morphology Dependent Double Yielding in Injection Molded Polycarbonate/Polyethylene Blend. *Macromol. Mater. Eng.* **2004,** *289,* 1004–1011.
74. Nakason, C.; Nuansomsri, K.; Kaesaman, A.; Kiatkamjornwong, S. Dynamic Vulcanization of Natural Rubber/High-Density Polyethylene Blends: Effect of Compatibilization, Blend Ratio and Curing System. *Polym. Test.* **2006,** *25,* 782–796.
75. Abreu, F. O. M. S.; Forte, M. M. C.; Liberman. S. A. SBS and SEBS Block Copolymers as Impact Modifiers for Polypropylene Compounds. *J. Appl. Polym. Sci.* **2005,** *95,* 254–263.
76. Jain, A. K.; Nagpal, A. K.; Singhal, R.; Gupta, N. K. Effect of Dynamic Crosslinking on Impact Strength and other Mechanical Properties of Polypropylene/Ethylene-Propylene-Diene Rubber Blends. *J. Appl. Polym. Sci.* **2000,** *78,* 2089–2103.
77. Nader, G.; Nouri, M. R.; Mehrabzadeh, M.; Bakhshandeh, G. R. Studies on Dynamic Vulcanization of PP/NBR Thermoplastic Elastomer Blends. *Iran Polym. J.* **1999,** *8,* 37–42.

78. Nakason, C.; Wannavilai, P.; Kaesaman, A. Effect of Vulcanization System on Properties of Thermoplastic Vulcanizates Based on Epoxidized Natural Rubber/Polypropylene Blends. *Polym. Test.* **2006,** *25,* 34–41.
79. Ismail, H.; Suryadiansyah. Thermoplastic Elastomers Based on Polypropylene/ Natural Rubber and Polypropylene/Recycle Rubber Blends. *Polym. Test.* **2002,** *21,* 389–395.
80. Bucknall, C. B. *Toughened Plastics;* Applied Science Publishers: London, 1977.
81. Paul, D. R.; Newman, S. *Polymer Blends;* Academic Press: New York, 1978.
82. Kinloch, A. J.; Young, R. J. *Fracture Behaviour of Polymers;* Applied Science Publishers: London, 1983.
83. Nakason, C.; Jarnthong, M.; Kaesaman, A.; Kiatkamjornwong, S. Influences of Blend Proportions and Curing Systems on Dynamic, Mechanical, and Morphological Properties of Dynamically Cured Epoxidized Natural Rubber/High-Density Polyethylene Blends. *Polym. Eng. Sci.* **2009,** *49,* 281–292.
84. Mohamed, M. G.; Messieh, S. L.; Sabbagh, S. E.; Younan, A. F. Electrical and Mechanical Properties of Polyethylene-Rubber Blends. *J. Appl. Polym. Sci.* **1998,** *69,* 775–783.
85. Yin, L.; Yin, J.; Shi, D.; Luan, S. Effects of SEBS-g-BTAI on the Morphology, Structure and Mechanical Properties of PA6/SEBS Blends. *Eur. Polym. J.* **2009,** *45,* 1554–1560.
86. Nakason, C.; Jamjinno, S.; Kaesaman, A.; Kiatkamjornwong, S. Thermoplastic Elastomer Based on High-Density Polyethylene/Natural Rubber Blends: Rheological, Thermal, and Morphological Properties. *Polym. Adv. Technol.* **2008,** *19,* 85–98.
87. Wu, S. Phase Structure and Adhesion in Polymer Blends: A Criterion for Rubber Toughening. *Polymer.* **1985,** *26,* 1855–1863.
88. Gaymans, R. J.; Borggreve, R. J. M.; Spoelstra, A. B. Ductile Transition in Nylon-Rubber Blends: Influence of Water. *J. Appl. Polym. Sci.* **1989,** *37,* 479–486.
89. Dijkstra, K.; Wevers, H. H.; Gaymans, R. J. Nylon-6/Rubber Blends: 7. Temperature-Time Effects in the Impact Behaviour of Nylon/Rubber Blends. *Polymer.* **1994,** *35,* 323–331.
90. Seo, Y.; Hwang, S. S.; Kim, K. U. Influence of the Mechanical Properties of the Dispersed Phase Upon the Behaviour of Nylon/Rubber Blends: Crosslinking Effect. *Polymer.* **1993,** *34,*1667–1676.
91. Borggreve, R. J. M.; Gaymans, R. J.; Eichenwald, H. M. Impact Behaviour of Nylon-Rubber Blends: 6. Influence of Structure on Voiding Processes; Toughening Mechanism. *Polymer.* **1989,** *30,* 78–83.
92. Borggreve, R. J. M.; Gaymans, R. J.; Schuijer, J. Impact Behaviour of Nylon-Rubber Blends: 5. Influence of the Mechanical Properties of the Elastomer. *Polymer.* **1989,** *30,* 71–77.
93. Wu, D.; Wang, X.; Jin, R. Effect of Nylon 6 on Fracture Behavior and Morphology of Tough Blends of Poly(2, 6-Dimethyl-1,4-Phenylene Oxide) and Maleated Styrene-Ethylene-Butadiene-Styrene Block Copolymer. *J. Appl. Polym. Sci.* **2006,** *99,* 3336–3343.
94. Kayano, Y.; Keskkula, H.; Paul, D. R. Evaluation of the Fracture Behaviour of Nylon 6/SEBS-g-MA Blends. *Polymer.* **1997,** *38,* 1885–1902.

A Review on Mechanical Properties of Semicrystalline

95. Huang, H.; Yang, J.; Xin Liu, X.; Zhang, Y. Dynamically Vulcanized Ethylene Propylene Diene Terpolymer/Nylon Thermoplastic Elastomers. *Eur. Polym. J.* **2002**, *38*, 857–861.

96. Tanrattanakul, V.; Sungthong, N.; Raksa, P. Rubber Toughening of Nylon 6 with Epoxidized Natural Rubber. *Polym. Test.* **2008**, *27*, 794–800.

97. Singh, H.; Gupta, N. K. Evolution of Properties in ABS/PA6 Blends Compatibilized by Fixed Weight Ratio SAGMA Copolymer. *J. Polym. Res.* **2011**, *18*, 1365–1377.

98. Xue, M. L.; Yu, Y.L.; Sheng, J.; Chuah, H. H.; Geng C. H. Compatibilization of Poly(Trimethylene Terephthalate)/Polycarbonate Blends by Epoxy. Part 1. Miscibility and Morphology. *J. Macromol. Sci. Phys.* **2005**, *44*, 317–329.

99. Yang, M.; Li, Z.; Feng, Studies on High Density Polyethylene/Polycarbonate Blend System Compatibilized with Low Density Polyethylene Grafted Diallyl Bisphenol A Ether. *J. Polym. Eng. Sci.* **1998**, *38*, 879–883.

100. Zhang, X. R.; Zhang, S. L.; Zhang, Y. N.; QIN, H. Y.; Jiang. D. Study on Compatibilized Polyethersulfone and Polycarbonate Blends. *J. Macromol. Sci. Phys.* **2011**, *50*, 1890–1904.

101. Zhihui, Y.; Yajie, Z.; Xiaomin, Z.; Jinghua, Y. Effects of the Compatibilizer PP-*g*-GMA on Morphology and Mechanical Properties of PP/PC Blends. *Polymer.* **1998**, *39*, 547–551.

102. Mbarek, S.; Jaziri, M. Recycling Poly(Ethylene Terephtalate) Wastes: Properties of Poly(Ethylene Terephtalate)/Polycarbonate Blends and the Effect of a Transesterification Catalyst. *Polym. Eng. Sci.* **2006**, *46*, 1378–1386.

103. Leclair, A.; Favis, B. D. The Role of Interfacial Contact in Immiscible Binary Polymer Blends and its Influence on Mechanical Properties. *Polymer.* **1996**, *37*, 4723–4728.

104. Pesetskii, S. S.; Jurkowski, B.; Koval V.N. Polycarbonate/Polyalkylene Terephthalate Blends: Interphase Interactions and Impact Strength. *J. Appl. Polym. Sci.* **2001**, *84*, 1277–1285.

105. Maier, C.; Calafut, T. *Polypropylene: The Definitive User's Guide and Data Book;* Elsevier Science: Amsterdam, Netherlands, 1998.

106. Kudva, R. A.; Keskkula, H.; Paul, D. R. Properties of Compatibilized nylon 6/ABS blends part I. Effect of ABS type. *Polymer.* **2000**, *41*, 225–237.

107. Tortorella, N.; Beatty, C. L. Morphology and Mechanical Properties of Impact Modified Polypropylene Blends. *Polym. Eng. Sci.* **2008**, *48*, 2098–2110.

108. Wang, X. H.; Zhang, H. X.; Jiang, W.; Wang Z. G.; Liu, C.H.; Liang, H. J.; Jiang, B. Z. Toughening of Nylon with Epoxidised Ethylene Propylene Diene Rubber. *Polymer,* **1998**, *39*, 2697–2699.

CHAPTER 2

PREPARATION OF POLYSACCHARIDE-BASED COMPOSITE MATERIALS USING IONIC LIQUIDS

JUN-ICHI KADOKAWA

Graduate School of Science and Engineering, Kagoshima University, 1-21-40 Korimoto, Kagoshima 8900065, Japan

*Corresponding author.E-mail: kadokawa@eng.kagoshima-u.ac.jp

CONTENTS

Abstract 34
2.1 Introduction 34
2.2 Preparation of Cellulose-Polymeric Ionic Liquid
     Composite Materials 38
2.3 Preparation of Chitin Nanofibrous Composite
     Materials Using Ionic Liquid 43
2.4 Conclusions 48
Acknowledgments 49
Keywords 49
References 50

ABSTRACT

This chapter reviews the preparation of polysaccharide-based composite materials using ionic liquids. A first topic deals with the fabrication of cellulose-polymeric ionic liquid composite materials by in situ polymerization method. Imidazolium-type ionic liquids having polymerizable groups have been used on the basis of concept on their properties that show a good affinity with cellulose, which suitably lead to well-compatibilization. Followings are the typical procedures for yielding the desired composites. Cellulose was first dissolved in ionic liquid solvents, followed by adding polymerizable ionic liquids or simply swollen with polymerizable ionic liquids. The systems were then heated in the presence of a radical initiator for the progress of in situ polymerization to give cellulose-polymeric ionic liquid composite materials. Compatibility or miscibility between cellulose and polymeric ionic liquids in the composites were evaluated by proper analytical measurements such as powder X-ray diffraction, thermal gravimetric analysis, and scanning electron microscopy. The mechanical properties were evaluated by tensile testing. As a second topic, the preparation of chitin nanofiber-synthetic polymer composite materials through the gelation of chitin with an ionic liquid was described. The author found that an ionic liquid, 1-allyl-3-methylimidazolium bromide (AMIMBr), dissolved chitin in concentrations up to ~4.8 wt% and mixtures of the higher amounts of chitin with AMIMBr gave ion gels. The chitin nanofibers were formed in the dispersion, which was produced by regeneration technique from the ion gel using methanol, followed by sonication. Moreover, filtration of the chitin nanofiber dispersion was carried out to give a chitin nanofiber film. The chitin nanofiber-synthetic polymer composite films were prepared by co-regeneration method and surface-initiated graft polymerization approach. By the former method, chitin nanofiber-poly(vinyl alcohol) composite film was obtained and the latter approach gave chitin nanofiber-poly(L-lactide-*co*-ε-caprolactone) composite film. The resulting composites can be expected as new bio-based functional materials to be used in different practical applications.

2.1 INTRODUCTION

Naturally occurring polysaccharides are widely distributed in nature and have increasingly been important because of their unique structures and properties.[1] Of the many kinds of polysaccharides, cellulose and chitin are the most important biomass resources[2,3] because they are the first and second most abundant natural polysaccharides on the earth, respectively. Cellulose is a glucose polymer consisting of β-(1→4)-linked glucose repeating units (Fig. 2.1(a)).[4] Cellulose has been studied on its fundamental chemical and physical properties and used in practical applications such as making cloths, furniture, and cosmetics. Chitin is a structurally similar polysaccharide as cellulose, but which has acetamido groups at the C-2 position in place of hydroxy groups of the glucose residues in cellulose (Fig. 2.1(b)).[5–7] Because cellulose and chitin are the abundant natural polysaccharides as aforementioned, there is major interest in their conversion into various useful materials after proper dissolution in suitable solvents. However, such polysaccharides as cellulose and chitin have often exhibited processability problems due to numerous hydrogen bonds in their polymeric chains, causing difficulty in employing them in a wide variety of new materials. Therefore, considerable efforts have been still devoted to compatibilization of such polysaccharides with synthetic polymers for improvement of their processability.

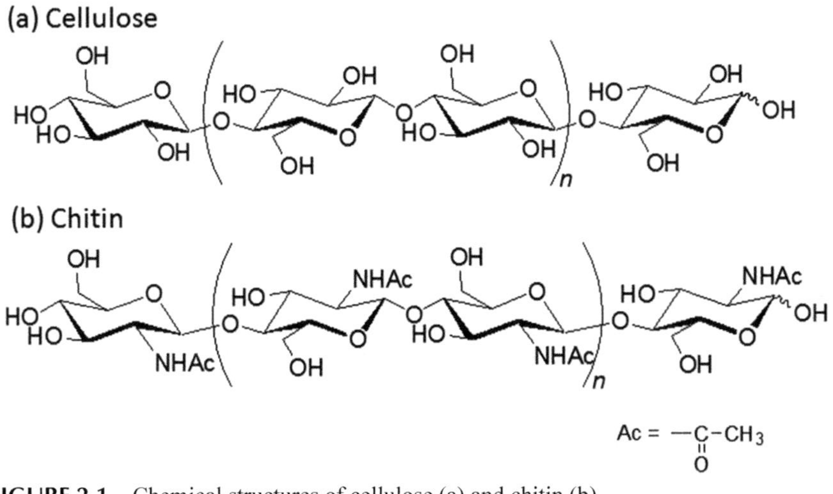

FIGURE 2.1   Chemical structures of cellulose (a) and chitin (b).

Ionic liquids (ILs), low-melting point salts that form liquids at temperatures below the boiling point of water,[8,9] have been found to act as good solvents for polysaccharides such as cellulose in the past decade. In 1934, it had already been discovered that a molten *N*-ethylpyridinium chloride, in the presence of nitrogen-containing bases, dissolved cellulose.[10] Although this was probably the first example of the cellulose dissolution using IL-type solvents, this was considered to be of less practical value at that time because the concept of ILs had not been put forward. In 2002, Rogers and co-workers comprehensively reported that an IL, 1-butyl-3-methylimidazolium chloride (BMIMCl), dissolved cellulose and this research opened up a new way for the development of a class of cellulose solvent system.[11] Since this publication, ILs, for example, mostly imidazolium-type ILs, have begun to be used in the processing of cellulose, which mainly concern with the dissolution, homogeneous derivatization and modification, and regeneration.[12–16] On the other hand, only the limited investigations have been reported regarding the dissolution of chitin with ionic liquids including the author's study,[17–19] in which the author found that an ionic liquid, 1-allyl-3-methylimidazolium bromide (AMIMBr), dissolved or swelled chitin to form weak gel-like materials (ion gels).[20,21]

Besides the use as solvents for polysaccharides, ILs have increasingly attracted much interesting attention as high-performance components with controllable physical and chemical properties as well as specific functions in composite materials.[22] To provide new polysaccharide-based composite materials composed of IL components, the author has focused on polymeric ILs (PILs, also often called polymerized ILs or polyILs), that is, the polymeric forms of ILs, which are prepared by polymerization of ILs having polymerizable groups (polymerizable ILs).[23,24] Because PILs have been studied on various applications such as membrane, porous and particle materials, electrolyte, sorbent, liquid crystal, and battery, furthermore, PILs can be expected to exhibit useful functions in the composites with polysaccharides. The target composite materials were facilely prepared by in situ radical polymerization of the imidazolium-type polymerizable ILs in the presence of polysaccharides.[25] The concept in this research topic is based on the properties of the imidazolium-type ILs that exhibit good affinities with the polysaccharides, which suitably lead to well-compatibilization.

The major advantages for employing PILs as material components are to improve processability and feasibility in application as practical materials. The imidazolium-type polymerizable ILs as a source of PIL components in the composites with polysaccharides are available by incorporating the polymerizable groups at anionic or cationic site in ILs.

Vinyl (meth)acryloyl and vinyl benzyl groups have typically been incorporated as the polymerizable group. Because polymerizable ILs having two polymerizable groups, such as acryl and vinyl groups, can be converted into insoluble and stable PILs with the cross-linked structure by the radical polymerization (Fig. 2.2), leading to enhanced stability and improved flexibility and durability, they have a highly potential as the source of the functional components in the practical materials.

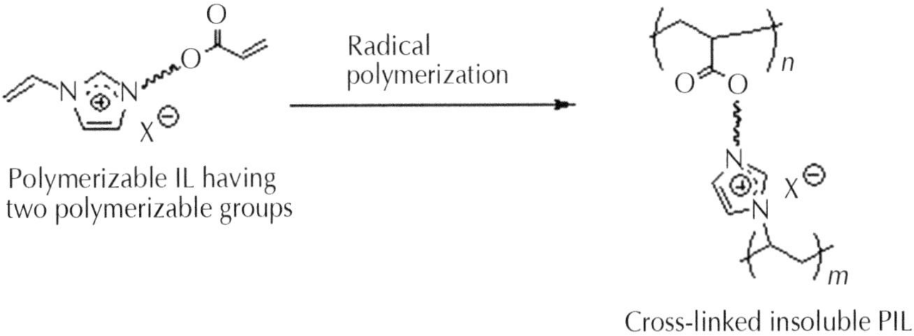

FIGURE 2.2   Polymerization of a polymerizable IL having two polymerizable groups to produce a cross-linked insoluble PIL.

On the other hand, preparation of nanofibrous polymeric assemblies is one of the most useful methods to practically utilize polymeric materials as observed in the case of cellulose and chitin.[26–29] For example, self-assembled fibrillar nanostructures from cellulose are promising materials for the practical applications in bio-related research fields such as tissue engineering. The efficient methods have also been developed for the preparation of chitin nanofibers. Conventional approaches to the production of chitin nanofibers are mainly performed upon top-down procedures that break down the starting bulk materials from native chitin resources (Fig. 2.3).[30–37] The other method accords to self-assembling generative (bottom-up) route, in which fibrillar nanostructures are produced by regeneration from chitin solutions via appropriate process (Fig. 2.3), as examples of the electrospinning equipment and simple precipitation process.[38–41] The author also reported that chitin nanofiber films were facilely obtained by regeneration from the aforementioned chitin ion gels with AMIMBr using methanol, followed by filtration according to the bottom-up procedure.[42]

On the basis of the above viewpoints and backgrounds, the first topic of this review deals with the efficient preparation of cellulose–PIL composite materials by in situ polymerization of polymerizable ionic liquids. As a second topic, the author describes the facile preparation of chitin nanofibers and the following fabrications of chitin-based nanofibrous composite materials with synthetic polymers using the ionic liquid.

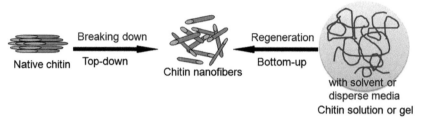

FIGURE 2.3 Top-down and bottom-up approaches for production of chitin nanofibers.

2.2 PREPARATION OF CELLULOSE-POLYMERIC IONIC LIQUID COMPOSITE MATERIALS

In situ polymerization method, for example, that is employed in interpenetrating polymer network (IPN) system, has been useful for the formation of composites from cellulose.[43–45] The IPNs are a special class of polymer composites based on two or more polymers, with each polymer chemically cross-linked or at least one network being prepared in the presence of the others.[46–48] Since the imidazolium-type ILs such as BMIMCl are good solvents for cellulose, it is interesting to employ the imidazolium-type PILs as one component in composite materials with cellulose.

To obtain the IPN-composites of cellulose with PIL by the in situ polymerization method, as the first study, a polymerizable IL, 1-(4-acryloyloxybutyl)-3-methylimidazolium bromide (AcMIMBr) was used[49] (Fig. 2.4). Cellulose was first dissolved with BMIMCl (9.1 wt%) by heating the mixture at 100°C for 24 h. Then, AcMIMBr (an equimolar amount of the glucose units in cellulose) and a radical initiator, AIBN, were added to the solution and the mixture was heated at 80°C for 5 h for the progress of radical polymerization. After the resulting mixture was washed with acetone and treated further with refluxed acetone, the residue was subjected to Soxhlet extraction with methanol to give the composite material. The unit ratio of cellulose to poly(AcMIMBr) was calculated by the elemental analysis to be 1.0:0.63 and the ^{1}H NMR spectrum of the

Soxhlet extract indicated that poly(AcMIMBr) was partially washed out by the extraction. The IR spectrum of the resulting material showed carbonyl absorption due to the ester linkage, supporting the composite structure composed of cellulose and poly(AcMIMBr). Furthermore, the X-ray diffraction (XRD) and thermal gravimetric analysis (TGA) data suggested the efficient compatibilization of cellulose with poly(AcMIMBr) in the composite. When cellulose and poly(AcMIMBr) were simply mixed in BMIMCl and the obtained mixture was subjected to the same isolation procedure as that for the aforementioned composite, poly(AcMIMBr) was mostly removed. This result indicated that the simple mixing of cellulose with poly(AcMIMBr) in BMIMCl was not efficient for the compatibilization.

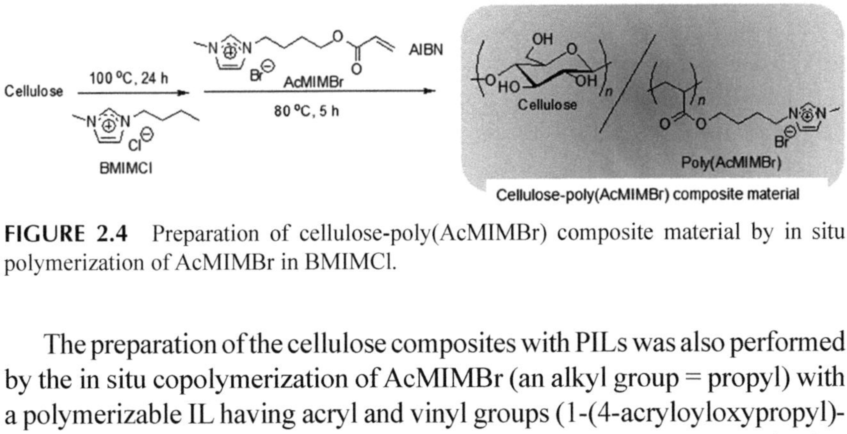

FIGURE 2.4 Preparation of cellulose-poly(AcMIMBr) composite material by in situ polymerization of AcMIMBr in BMIMCl.

The preparation of the cellulose composites with PILs was also performed by the in situ copolymerization of AcMIMBr (an alkyl group = propyl) with a polymerizable IL having acryl and vinyl groups (1-(4-acryloyloxypropyl)-3-vinylimidazolium bromide, AcVIMBr) (Fig. 2.5).[50] The copolymerization of these polymerizable ILs gives cross-linked insoluble poly(AcMIMBr-co-AcVIMBr) because of the presence of two polymerizable groups in AcVIMBr. The properties of poly(AcMIMBr-co-AcVIMBr)s depend on cross-linked densities, which are controlled by the feed ratios of the two polymerizable ILs in the copolymerization. Furthermore, it was found that these polymerizable ILs had the ability to largely disrupt the crystalline structure of cellulose by immersion of cellulose into them, indicating that the composites were obtained by the procedure without the use of the solvent IL such as BMIMCl unlike the case in the above study.

Thus, the preparation of the cellulose composites using AcMIMBr and AcVIMBr was conducted as follows. Cellulose (30 wt% for AcMIMBr + AcVIMBr) was first immersed in mixtures of AcMIMBr

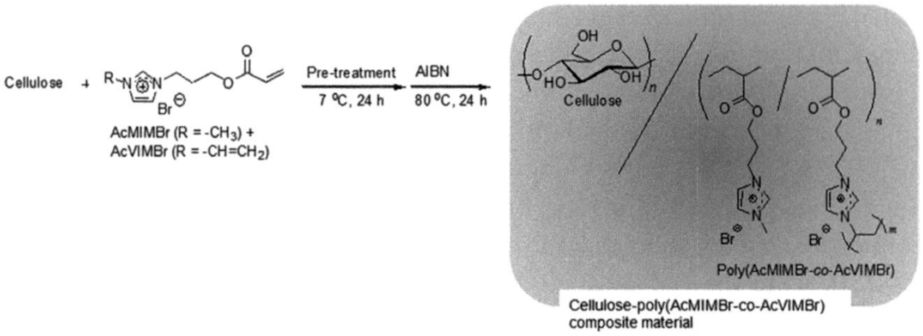

FIGURE 2.5 Preparation of cellulose-poly(AcMIMBr-*co*-AcVIMBr) composite material by in situ copolymerization of AcMIMBr with AcVIMBr.

and AcVIMBr with various weight ratios (100:0–50:50) at 7°C for 24 h. The XRD profile of the resulting pre-treated system (AcMIMBr/ AcVIMBr = 95/5) showed large diminution of the crystalline peaks of cellulose (Fig. 2.6(a) and (b)). The TGA result of the pre-treated system exhibited an onset weight loss at around 250°C due to the thermal degradation of cellulose, which was ca. 50°C lower than that of the original cellulose. The XRD and TGA data indicated that the cellulose chains were swollen in the pre-treated system using the mixtures of two polymerizable ILs. After AIBN was added to the pre-treated mixtures, then, the system was heated at 80°C for 24 h for the progress of the in situ radical copolymerization. Consequently, the composites were obtained without any isolation and purification procedures.

The compatibility between cellulose and PILs in the composites was evaluated by the scanning electron microscopic (SEM) measurement. The SEM image of the composite (AcMIMBr/AcVIMBr = 95/5) showed the completely different morphology from that of the original cellulose, which indicated the good compatibility between two polymeric components in the composite (Fig. 2.7(a) and (b)). The stability of the composites was examined by washing with methanol using the Soxhlet extraction manner. After washing the composites obtained by using AcMIMBr/AcVIMBr (95/5), the unit ratio of cellulose to PILs was not mostly changed from that before washing. On the other hand, the content of the PIL in the composite resulted by using a sole AcMIMBr decreased by washing. The results indicated that the composite from AcMIMBr/AcVIMBr was quite stable owing to consisting of the cross-linked poly(AcMIMBr-*co*-AcVIMBr). In contrast, the PIL component (poly(AcMIMBr)) in the composite from

AcMIMBr was washed out, due to its linear structure, which was soluble in methanol.

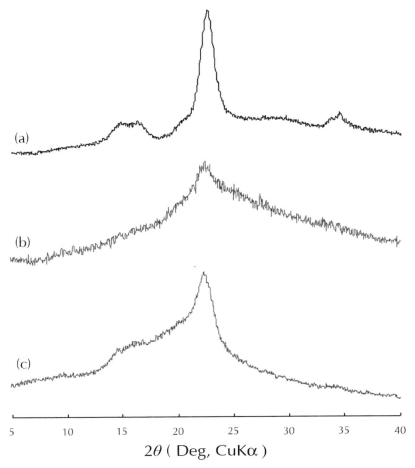

FIGURE 2.6 XRD profiles of cellulose (a) and pre-treated mixtures of cellulose with AcMIMBr and AcVIMBr (b) and with VVBnIMCl (c).

The in situ polymerization method was applied to the preparation of cellulose-based ionic porous material compatibilized with PIL.[51] This was achieved by combining the in situ polymerization method with the templating technique using the oil/IL emulsion system (Fig. 2.8).[52] In situ copolymerization of AcMIMBr with AcVIMBr was first performed in a solution of cellulose with a solvent of BMIMCl. Then, the sonication of the mixture coexisting with corn oil and sorbitan monooleate was done, followed by the appropriate treatment procedures to give porous material. The pore sizes of the material ascertained from the SEM image were found to be around 0.15–1.3 μm accompanied with some smaller pores with size ranging from 30 to 70 nm (Fig. 2.7(c)).

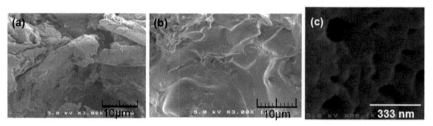

FIGURE 2.7 SEM images of cellulose (a), cross-sectional area of cellulose-poly(AcMIMBr-co-AcVIMBr) composite material (b), and cellulose-based ionic porous material compatibilized with poly(AcMIMBr-co-AcVIMBr) (c).

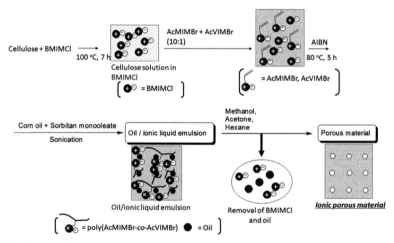

FIGURE 2.8 Schematic image for preparation of cellulose-based ionic porous material compatibilized with poly(AcMIMBr-co-AcVIMBr).

The in situ polymerization method was extended to use another polymerizable IL, which was an IL having vinylbenzyl and vinyl polymerizable groups (1-vinyl-3-(4-vinylbenzyl)imidazolium chloride, VVBnIMCl) (Fig. 2.9).[53] Cellulose (9.1–50.0 wt%) was first pre-treated with VVBnIMCl by immersing cellulose into VVBnIMCl at 7°C for 24 h. The TGA result of

Preparation of Polysaccharide-Based Composite Materials 43

the pre-treated system (33.3 wt%) showed the similar profile as that using AcMIMBr/AcVIMBr. However, the XRD result of the pre-treated system (Fig. 2.6(c)) in comparison with that using AcMIMBr/AcVIMBr indicated that the ability of AcMIMBr/AcVIMBr disrupting the crystalline structure of cellulose was much higher than that of VVBnIMCl. After AIBN was added to the pre-treated mixtures, then, the system was heated at 80°C for 24 h, giving rise to the composites without any further isolation and purification procedures. The XRD and TGA results of all the composites indicated the partial disruption of the crystalline structure of cellulose.

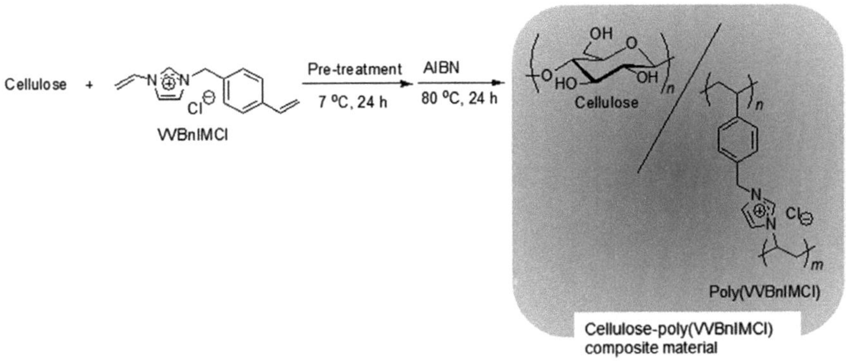

FIGURE 2.9 Preparation of cellulose-poly(VVBnIMCl) composite material by in situ polymerization of VVBnIMCl.

2.3 PREPARATION OF CHITIN NANOFIBROUS COMPOSITE MATERIALS USING IONIC LIQUID

In addition to ionic liquids, 1-butyl- and 1-ethyl-3-methylimidazolium acetates, that dissolve chitin,[17,19] for the further dissolution study of chitin with ILs, the author noted the previous study reporting that the imidazolium-type ionic liquids having a bromide counter anion were good solvents for the synthesis of polyamides and polyimides.[54] The result inspired to use the same kind of ILs for dissolution of chitin, because chitin has the –N–C=O groups, same as polyamides and polyimides. Consequently, it was found that AMIMBr dissolved chitin in the concentrations up to ~4.8 wt% when the dissolution experiments using several imidazolium bromide ionic liquids were conducted (Fig. 2.10(a)).[20,21] Interestingly, mixtures of higher amounts of chitin with AMIMBr gave more viscous materials, that is, the ion gels when 6.5–11 wt% of chitin were immersed in AMIMBr at room temperature for 24 h, followed by heating at 100°C for 48 h and cooling to room temperature.[20] Indeed, the resulting 6.5 wt% chitin with AMIMBr did not flow upon leaning a test tube (Fig. 2.10(b)), whereas the aforementioned 4.8 wt% chitin with AMIMBr started to flow upon leaning (Fig. 2.10(a)).

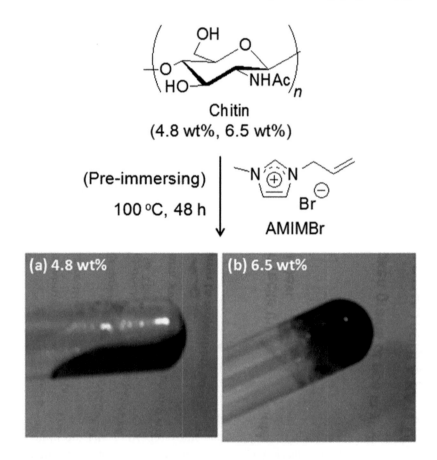

FIGURE 2.10 Dissolution (4.8 wt%, a) and gelation (6.5 wt%, b) of chitin with AMIMBr.

As mentioned above, the author has reported that chitin nanofiber films were facilely obtained by regeneration from the chitin ion gels with AMIMBr using methanol, followed by filtration (Fig. 2.11).[42] First, chitin was swollen with AMIMBr according to the procedure reported in the previous study as aforementioned to give the chitin ion gels with AMIMBr (9.1–11 wt%). It was found that chitin dispersions were obtained when the gels were treated with methanol at room temperature for 24 h to slowly regenerate chitin, followed by sonication. The resulting dispersion was diluted with methanol, which was subjected to the SEM measurement. The morphology of nanofibers with ca. 20–60 nm in width and several hundred nanometers in length was seen in the SEM image of the sample from the dispersion (Fig. 2.11), indicating the formation of the chitin nanofibers by the above gelation and regeneration procedures of chitin. When the dispersion was filtered, the residue formed a film, which was further purified by Soxhlet extraction with methanol. The SEM image of the resulting film was also measured to confirm the nano-scaled morphology of chitin (Fig. 2.11), which showed the pattern of highly entangled nanofibers. Such entangled structure of the nanofibers probably contributed to formability of the film. The XRD pattern of the chitin nanofiber film was identical with that of the original chitin powder, indicating that the crystalline structure was reconstructed by the above regeneration procedure during the formation of the nanofibers.

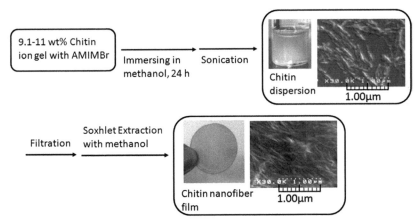

FIGURE 2.11 Procedures for preparation of chitin dispersion and nanofiber film by regeneration from chitin ion gel with AMIMBr.

As one of the possible applications of the present chitin nanofiber film, attempts were made to prepare the chitin nanofiber composite materials with synthetic polymers. Two kinds of approaches, that is, physical and chemical approaches, have been considered to yield the chitin-synthetic polymer composite materials (Fig. 2.12). In former case, the chitin and synthetic polymer chains construct material components by physical interaction in the composites, whereas the latter approach results in the formation of covalent linkages between two polymer chains in the composites.

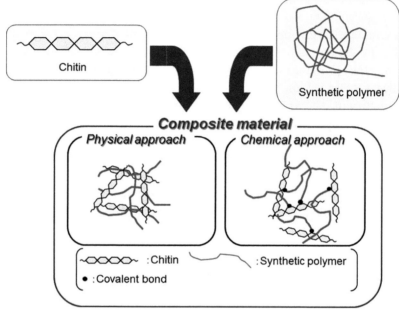

FIGURE 2.12 Chitin-synthetic polymer composite materials by physical and chemical approaches.

By the physical approach, the chitin nanofiber-poly(vinyl alcohol) (PVA) composite film was prepared (Fig. 2.13).[42] First, the 9.1 wt% chitin ion gel with AMIMBr was prepared according to the aforementioned procedure and a solution of PVA (DP = ca. 4300) in hot water was added to the gel (the feed weight ratio of chitin to PVA = 1 : 0.30). Then, the co-regeneration, filtration, and Soxhlet extraction procedures gave the chitin nanofiber–PVA composite film. The SEM image of the composite showed that the nanofiber-like morphology was maintained and PVA components probably filled in spaces among the fibers (Fig. 2.13), indicating relative immiscibility of chitin and PVA in the composite. The DSC profile of the composite film exhibited an endothermic peak due to the melting point of PVA, but which was broadened, indicating that the crystallinity of PVA decreased in the composite film.

Preparation of Polysaccharide-Based Composite Materials

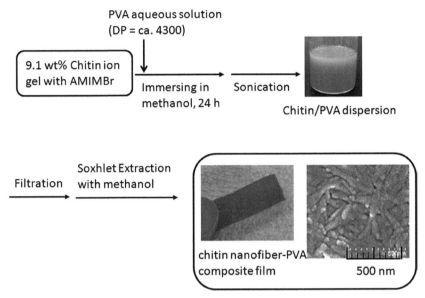

FIGURE 2.13 Procedure for preparation of chitin nanofiber–PVA composite film.

The biodegradable polyesters are efficiently synthesized by ring-opening polymerization of the corresponding cyclic ester monomers, such as L-lactide (LA) and ε-caprolactone (CL), initiated with hydroxy groups in the presence of Lewis acid catalysts such as tin(II) 2-ethylhexanoate.[55,56] To incorporate the biodegradable polyesters on the aforementioned chitin nanofiber films, surface-initiated ring-opening graft copolymerization of LA/CL monomers initiated from hydroxy groups on the chitin nanofiber was attempted to produce chitin nanofiber-*graft*-poly(L-lactide-*co*-ε-caprolactone) (chitin nanofiber-g-poly(LA-*co*-CL)) films (Fig. 2.14).[57] To efficiently initiate the ring-opening graft copolymerization of LA/CL on surface of the nanofibers, spaces among the fibers were made by immersing the film in water for 10 s. Then, the surface-initiated ring-opening graft copolymerization of LA/CL (feed molar ratio = 20:80) from the pre-treated film was carried out. First, the film was immersed in a solution of LA/CL in toluene. Then, the polymerization catalyst, tin(II) 2-ethylhexanoate (*ca.* 5 mol% for the monomers), was added and the mixture was heated at 80°C for 48 h for the progress of the ring-opening copolymerization. After the resulting film was washed with chloroform and further subjected to Soxhlet extraction with chloroform, it was dried under reduced pressure to give chitin nanofiber-g-poly(LA-co-CL) film. The IR spectrum of the resulting film exhibited carbonyl absorption at 1739 cm^{-1} assignable to the ester linkage of poly(LA-co-CL), suggesting the presence of poly(LA-co-CL) in the product, which was explanatorily bound to the nanofibers by covalent linkage. The LA/CL compositional ratio of the grafted poly(LA-co-CL) (40/60) was higher than that in the feed (20/80) because of the higher reactivity of LA than CL in the copolymerization.[58]

The SEM image of the resulting chitin nanofiber-g-poly(LA-co-CL) film showed that the nanofibers still remained (Fig. 2.14), but the fiber widths increased (60–100 nm) compared with those of the original chitin nanofiber film and some fibers were merged at the interfacial areas, that was probably caused by the grafted polyesters present on the nanofibers. The XRD profile of the chitin nanofiber-g-poly(LA-co-CL) film exhibited the same pattern as that of the original film, indicating that the crystalline structure of the chitin chains was not disrupted owing to the occurrence of the graft copolymerization only on surface of the nanofibers.

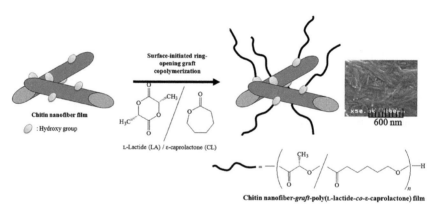

FIGURE 2.14 Surface-initiated ring-opening copolymerization of LA/CL from chitin nanofiber film.

2.4 CONCLUSIONS

This chapter overviewed the preparation of polysaccharide-based composite materials using ionic liquids. On the basis of the viewpoint that the imidazolium-type ILs show good affinity with cellulose, the well-compatible materials of cellulose with the imidazolium-type PILs have been produced by the in situ polymerization technique. The author also overviewed the preparation of chitin nanofiber–synthetic polymer composite materials through the gelation with AMIMBr. First, the preparation of chitin nanofiber film was performed by regeneration from the chitin ion gel with AMIMBr. The composite film of the chitin nanofibers with PVA was prepared according to the co-regeneration technique by a similar procedure as that for the chitin nanofiber film. Moreover, the preparation of chitin nanofiber-*g*-poly(LA-*co*-CL) films was conducted by the surface-initiated ring-opening graft copolymerization of LA/CL monomers from the chitin nanofiber film.

This type of the research described herein provides new polysaccharide-based functional materials, which have a high potential for practical applications. This review has presented the approach of how the polysaccharide-based composite materials were facilely obtained using ionic liquids, which will be increasingly developed with attracting much attention depending not only on the further technological interests on this type of the composites, but also on the scientific curiosity of the scientists.

ACKNOWLEDGMENTS

The author is indebted to the co-workers, whose names are found in references from his papers, for their enthusiastic collaborations.

KEYWORDS

- cellulose
- chitin
- composite
- ionic liquid
- nanofibers
- polysaccharide

REFERENCES

1. Schuerch, C. Polysaccharides. In *Encyclopedia of Polymer Science and Engineering*, 2nd ed.; Mark, H. F.; Bilkales, N.; Overberger, C. G., Eds.; John Wiley & Sons: New York, 1986; Vol. 13; pp 87–162.
2. Mohanty, A. K.; Misra, M.; Drzal, L. T. Sustainable Bio-Composites from Renewable Resources: Opportunities and Challenges in the Green Materials World. *J. Polym. Environ.* **2002,** *10,* 19–26.
3. Rouilly, A.; Rigal, L. Agro-Materials: A Bibliographic Review. *J. Macromol. Sci. Polym. Rev.* **2002,** *42,* 441–479.
4. Klemm, D.; Heublein, B.; Fink, H. -P.; Bohn, A. Cellulose: Fascinating Biopolymer and Sustainable Raw Material. *Angew. Chem. Int. Ed.* **2005,** *44,* 3358–3393.
5. Kurita, K. Chitin and Chitosan: Functional Biopolymers from Marine Crustaceans. *Marine Biotech.* **2006,** *8,* 203–226.
6. Pillai, C. K. S.; Paul, W.; Sharma, C. P. Chitin and Chitosan Polymers: Chemistry, Solubility and Fiber Formation. *Prog. Polym. Sci.* **2009,** *34,* 641–678.
7. Rinaudo, M. Chitin and Chitosan: Properties and Applications. *Prog. Polym. Sci.* **2006,** *31,* 603–632.
8. Wasserscheid, P.; Keim, W. Ionic Liquids-New "Solutions" for Transition Metal Catalysis. *Angew. Chem. Int. Ed.* **2000,** *39,* 3772–3789.
9. Welton, T. Room-Temperature Ionic Liquids. Solvents for Synthesis and Catalysis. *Chem. Rev.* **1999,** *99,* 2071–2084.
10. Graenacher, C. Cellulose Solution. U.S. Patent 1,943,176, January 9, 1934.
11. Swatloski, R. P.; Spear, S. K.; Holbrey, J. D.; Rogers, R. D. Dissolution of Cellulose with Ionic Liquids. *J. Am. Chem. Soc.* **2002,** *124,* 4974–4975.
12. Feng, L.; Chen, Z. I. J. Research Progress on Dissolution and Functional Modification of Cellulose in Ionic Liquids. *Mol. Liq.* **2008,** *142,* 1–5.
13. Liebert, T.; Heinze, T. Interaction of Ionic Liquids with Polysaccharides 5. Solvents and Reaction Media for the Modification of Cellulose. *Bioresources.* **2008,** *3,* 576–601.
14. Pinkert, A.; Marsh, K. N.; Pang, S.; Staiger, M. P. Ionic Liquids and Their Interaction with Cellulose. *Chem. Rev.* **2009,** *109,* 6712–6728.
15. Zakrzewska, M. E.; Lukasik, E. B.; Lukasik, R. B. Solubility of Carbohydrates in Ionic Liquids. *Energy Fuels.* **2010,** *24,* 737–745.
16. Seoud, O. A. E.; Koschella, A.; Fidale, L. C.; Dorn, S.; Heinze, T. Applications of Ionic Liquids in Carbohydrate Chemistry: A Windows of Opportunities.
17. Qin, Y.; Lu, X.; Sun, N.; Rogers, R. D. Dissolution or Extraction of Crustacean Shells Using Ionic Liquids to Obtain High Molecular Weight Purified Chitin and Direct Production of Chitin Films and Fibers. *Green Chem.* **2010,** *12,* 968–971.
18. Wang, W. T.; Zhu, J.; Wang, X. L.; Huang, Y.; Wang, Y. Z. Dissolution Behavior of Chitin in Ionic Liquids. *J. Macromol. Sci. Part B Phys.* **2010,** *49,* 528–541.
19. Wu, Y.; Sasaki, T.; Irie, S.; Sakurai, K. A Novel Biomass-Ionic Liquid Platform for the Utilization of Native Chitin. *Polymer.* **2008,** *49,* 2321–2327.
20. Prasad, K.; Murakami, M.; Kaneko, Y.; Takada, A.; Nakamura, Y.; Kadokawa, J. Weak Gel of Chitin with Ionic Liquid, 1-Allyl-3-Methylimidazolium Bromide. *Int. J. Biol. Macromol.* **2009,** *45,* 221–225.

21. Yamazaki, S.; Takegawa, A.; Kaneko, Y.; Kadokawa, J.; Yamagata, M.; Ishikawa, M. An Acidic Cellulose-Chitin Hybrid Gel as Novel Electrolyte for an Electric Double Layer Capacitor. *Electrochem. Commun.* 2009, *11*, 68–70.
22. Giernoth, R. Task-Specific Ionic Liquids. *Angew. Chem. Int. Ed.* **2010**, *49*, 2834–2839.
23. Green, O.; Grubjesic, S.; Lee, S.; Firestone, M. A. The Design of Polymeric Ionic Liquids for the Preparation of Functional Materials. *Polym. Rev.* **2009**, *49*, 339–360.
24. Mecerreyes, D. Polymeric Ionic Liquids: Broadening the Properties and Applications of Polyelectrolytes. *Prog. Polym. Sci.* **2011**, *36*, 1629–1648.
25. Kadokawa, J. Preparation of Polysaccharide-Based Materials Compatibilized with Ionic Liquids. In *Ionic Liquids, Application and Perspectives;* Kokorin, A., Ed.; InTech: Rijeka, Croatia, 2011; pp 95–114.
26. Abdul Khalil, H. P. S.; Bhat, A. H.; Ireana Yusra, A. F. Green Composites from Sustainable Cellulose Nanofibrils: A Review. *Carbohydr. Polym.* **2012**, *87*, 963–979.
27. Ifuku, S. Preparation of Chitin Nanofibers from Crab Shell and Their Applications. *Kobunshi Ronbunshu.* **2012**, *69*, 460–467.
28. Isogai, A.; Saito, T.; Fukuzumi, H. TEMPO-Oxidized Cellulose Nanofibers. *Nanoscale.* **2011**, *3*, 71–85.
29. Zeng, J. B.; He, Y. S.; Li, S. L.; Wang, Y. Z. Chitin Whiskers: An Overview. *Biomacromolecules.* **2012**, *13*, 1–11.
30. Fan, Y.; Saito, T.; Isogai, A. Chitin Nanocrystals Prepared by TEMPO-Mediated Oxidation of α-Chitin. *Biomacromolecules.* **2008**, *9*, 192–198.
31. Fan, Y.; Saito, T.; Isogai, A. TEMPO-Mediated Oxidation of β-Chitin to Prepare Individual Nanofibers. *Carbohydr. Polym.* **2009**, *77*, 832–838.
32. Goodrich, J. D.; Winter, W. T. α-Chitin Nanocrystals Prepared from Shrimp Shells and Their Specific Surface Area Measurement. *Biomacromolecules.* **2007**, *8*, 252–257.
33. Ifuku, S.; Nogi, M.; Abe, K.; Yoshioka, M.; Morimoto, M.; Saimoto, H.; Yano, H. Preparation of Chitin Nanofibers with a Uniform Width as α-Chitin from Crab Shells. *Biomacromolecules.* **2009**, *10*, 1584–1588.
34. Ifuku, S.; Nogi, M.; Yoshioka, M.; Morimoto, M.; Yano, H.; Saimoto, H. Fibrillation of Dried Chitin into 10-20 nm Nanofibers by a Simple Grinding Method under Acidic Conditions. *Carbohydr. Polym.* **2010**, *81*, 134–139.
35. Li, J.; Revol, J. F.; Naranjo, E.; Marchessault, R. H. Effect of Electrostatic Interaction on Phase Separation of Chitin Crystallite Suspensions. *Int. J. Biol. Macromol.* **1996**, *18*, 177–187.
36. Li, J.; Revol, J. F.; Marchessault, R. H. Effect of Degree of Deacetylation of Chitin on the Properties of Chitin Crystallites. *J. Appl. Polym. Sci.* **1997**, *65*, 373–380.
37. Revol, J. F.; Marchessault, R. H. In-Vitro Chiral Nematic Ordering of Chitin Crystallites. *Int. J. Biol. Macromol.* **1993**, *15*, 329–335.
38. Jayakumar, R.; Prabaharan, M.; Nair, S. V.; Tamura, H. Novel Chitin and Chitosan Nanofibers in Biomedical Applications. *Biotechnol. Adv.* **2010**, *28*, 142–150.
39. Schiffman, J. D.; Stulga, L. A.; Schauer, C. L. Chitin and Chitosan: Transformations Due to the Electrospinning Process. *Polym. Eng. Sci.* **2009**, *49*, 1918–1928. *Biomacromolecules.* **2007**, *8*, 2629–2648.

40. Zhong, C.; Cooper, A.; Kapetanovic, A.; Fang, Z.; Zhang, M.; Rolandi, M. A Facile Bottom-Up Route to Self-Assembled Biogenic Nanofibers. *Soft Matter.* **2010,** *6,* 5298–5301.

41. Zhong, C.; Kapetanovic, A.; Deng, Y.; Rolandi, M. A Chitin Nanofiber Ink for Airbrushing, Replica Molding, and Microcontact Printing of Self-Assembled Macro-, Micro-, and Nanostructures. *Adv. Mater.* **2011,** *23,* 4776–4781.

42. Kadokawa, J.; Takegawa, A.; Mine, S.; Prasad, K. Preparation of Chitin Nanowhiskers Using an Ionic Liquid and Their Composite Materials with Poly(Viny Alcohol). *Carbohydr. Polym.* **2011,** *84,* 1408–1412.

43. Nishio, Y. *Hyperfine Composites of Cellulose with Synthetic Polymers.* In *Cellulosic Polymers, Blends and Composites*; Gilbert, R. D., Ed.; Carl Hanser: Munich, 1994; Chap 5; pp 95–113.

44. Nishio, Y.; Hirose, N. Cellulose/Poly(2-Hydroxyethyl Methacrylate) Composites Prepared via Solution Coagulation and Subsequent Bulk Polymerization. *Polymer.* **1992,** *33,* 1519–1524.

45. Miyashita, Y.; Nishio, Y.; Kimura, N.; Suzuki, H.; Iwata, M. Transition Behaviour of Cellulose/Poly(*N*-Vinylpyrrolidone-*Co*-Glycidyl Methacrylate) Composites Synthesized by a Solution Coagulation/Bulk Polymerization Method. *Polymer.* **1996,** *37,* 1949–1957.

46. Klempuner, D.; Frisch, K. C. *Advances in Interpenetrating Polymer Networks*; Technomic Publishing Co: Lancaster, 1989; Vols. I–III.

47. Lipatov, Y. S.; Alekseeva, T. T. Phase-Separated Interpenetrating Polymer Networks. *Adv. Polym. Sci.* **2007,** *208,* 1–227.

48. Sperling, L. H. *Interpenetrating Polymer Networks and Related Materials;* Plenum Press: New York, 1981.

49. Murakami, M.; Kaneko, Y.; Kadokawa, J. Preparation of Cellulose-Polymerized Ionic Liquid Composite by in situ Polymerization of Polymerizable Ionic Liquid in Cellulose-Dissolving Solution. *Carbohydr. Polym.* **2007,** *69,* 378–381.

50. Takegawa, A.; Murakami, M.; Kaneko, Y.; Kadokawa, J. A Facile Preparation of Composites Composed of Cellulose and Polymeric Ionic Liquids by in situ Polymerization of Ionic Liquids Having Acrylate Groups. *Polym. Compos.* **2009,** *30,* 1837–1841.

51. Prasad, K.; Mine, S.; Kaneko, Y.; Kadokawa, J. Preparation of Cellulose-Based Ionic Porous Material Compatibilized with Polymeric Ionic Liquid. *Polym. Bull.* **2010,** *64,* 341–349.

52. Zhao, X. S.; Su, F.; Yan, Q.; Guo, W.; Bao, X. Y.; Lv, L.; Zhou, Z. Templating Methods for Preparation of Porous Structures. *J. Mater. Chem.* **2006,** *16,* 637–648.

53. Kadokawa, J.; Murakami, M.; Kaneko, Y. A Facile Method for Preparation of Composites Composed of Cellulose and a Polystyrene-Type Polymeric Ionic Liquid Using a Polymerizable Ionic Liquid. *Compos. Sci. Technol.* **2008,** *68,* 493–498.

54. Vygodskii, Y. S.; Lozinskaya, E. I.; Shaplov, A. S. Ionic Liquids as Novel Reaction Media for the Synthesis of Condensation Polymers. *Macromol. Rapid Commun.* **2002,** *23,* 676–680.

55. Jérôme, C.; Lecomte, P. Recent Advances in the Synthesis of Aliphatic Polyesters by Ring-Opening Polymerization. *Adv. Drug Deliv. Rev.* **2008,** *60,* 1056–1076.

56. Lou, X.; Detrembleur, C.; Jérôme, R. Novel Aliphatic Polyesters Based on Functional Cyclic (Di)Esters. *Macromol. Rapid Commun.* **2003,** *24,* 161–172.
57. Setoguchi, T.; Yamamoto, K.; Kadokawa, J. Preparation of Chitin Nanofiber-Graft-Poly(L-Lactide-*co*-ε-Caprolactone) Films By Surface-Initiated Ring-Opening Graft Copolymerization. *Polymer.* **2012,** *53,* 4977–4982.
58. Grijpma, D. W.; Pennings, A. J. Polymerization Temperature Effects on the Properties of L-lactide and ε-Caprolactone Copolymers. *Polym. Bull.* **1991,** *25,* 335–341.

CHAPTER 3

MICROCRYSTALLINE CELLULOSE: AN OVERVIEW

SRIMANTA SARKAR[1,2] CELINE VALERIA LIEW[1,*]
JOSEPHINE LAY PENG SOH[1,3] PAUL WAN SIA HENG[1]
and TIN WUI WONG[4,5*]

[1]*Department of Pharmacy, Faculty of Science, GEA-NUS Pharmaceutical Processing Research Laboratory, National University of Singapore, 18 Science Drive 4, Singapore 117543, Singapore*

[2]*Novartis Singapore Pharmaceutical Manufacturing Pte. Ltd., 10 Tuas Bay Lane, Singapore 637461, Singapore*

[3]*Pharma Tech Ops Manufacturing Science and Technology, Novartis Pharma AG, Basel, Switzerland*

[4]*Non-Destructive Biomedical and Pharmaceutical Research Centre, iPROMISE, Universiti Teknologi MARA, Puncak Alam 42300, Selangor, Malaysia*

[5]*Particle Design Research Group, Faculty of Pharmacy, Universiti Teknologi MARA, Puncak Alam 42300, Selangor, Malaysia*

Corresponding author. E-mail: phalcv@nus.edu.sg; wongtinwui@salam.uitm.edu.my

CONTENTS

Abstract 56
3.1 Introduction...56
3.2 Structure and Manufacture..57
3.3 Future Perspectives 70
Keywords 70
References..71

ABSTRACT

Microcrystalline cellulose (MCC) is obtained from naturally occurring cellulose that has been purified and partially depolymerized. MCC is an excipient with wide-ranging applications in the pharmaceutical industry. Particularly, in the process of extrusion–spheronization, MCC has shown unsurpassed efficiency in terms of process control and end-product quality. It has been regarded as an essential excipient for a well-controlled system for pellet production. Inherent variability in the physical properties of MCC types obtained from different manufacturing processes and suppliers has been shown to affect their function and performance, and consequently process control during extrusion–spheronization process and end-product quality. This chapter provides an outline of the different types of MCC obtained from different manufacturing processes as well as different suppliers, and the current state of knowledge on its application in extrusion–spheronization as a pelletization aid.

3.1 INTRODUCTION

Microcrystalline cellulose (MCC) is manufactured by purification and partial depolymerization of naturally occurring cellulose. MCC is used widely as an additive in the pharmaceutical, food, and cosmetic industries. In the pharmaceutical industry, it is a well-known multifunctional excipient and is used in the formulation of a variety of dosage forms as viscosity controlling agent, gelling agent, texture modifier, suspension stabilizer, water absorber, non-adhesive binder, emulsifier, disintegrant, filler, and pelletization aid. Due to its favorable compression characteristics, MCC was initially employed as a directly compressible excipient for tableting. However, it is the use of MCC in wet granulation, particularly for pellet production, that generated extensive research interest in the past few decades. It is generally accepted that a pelletization aid is an essential component in the formulation for extrusion–spheronization to yield highly spherical pellets with narrow size distribution. The most commonly used pelletization aid is MCC and it is considered as the gold standard.

The functionality of MCC is largely dependent on its physical properties which, in turn, affect process control and end-product quality attributes. Source/supplier variability in MCC grades which includes the

Microcrystalline Cellulose: An Overview 57

starting material and manufacturing process can be considered as one of the main contributors in the difference in MCC quality as reflected in the basic physicochemical properties such as particle size and size distribution, micromeritics, crystallinity, and packing properties, e.g., bulk and tapped densities.[1-7] This chapter provides an outline of the pharmaceutical applications of MCC using a range of MCC grades from different sources and manufacturing process to illustrate the effect of variability on different pharmaceutical processes, particularly on extrusion spheronization.

3.2 STRUCTURE AND MANUFACTURE

MCC is prepared from highly purified wood pulp by hydrolysis in acidic medium, usually a dilute mineral acid solution. Acid hydrolysis cleaves the β-1, 4 glucosidic linkages between the glucopyranose units, effecting the removal of the amorphous region. The formed water-soluble components, i.e., cello-oligosaccharides and glucose, are removed by repeated rinsing and filtration. The free MCC microcrystals or microfibrils (5–10 nm in width) obtained by subsequent neutralization and mechanical agitation of the aqueous slurry are composed of tight bundles of linear cellulose chains (typically 200–300 units long). These cellulose chains are held by hydrogen bonds via the hydroxyl moieties and aggregated into much larger crystallites (>60 % crystallinity) that are referred to as primary MCC particles.[6,8] The chemical structure of MCC is shown in Figure 3.1.

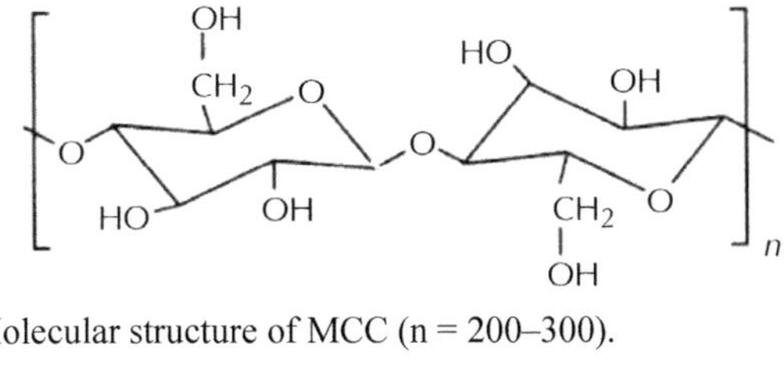

FIGURE 3.1   Molecular structure of MCC (n = 200–300).

3.2.1   AVAILABLE FORMS OF MCC

Three different types of MCC, namely powdered MCC, colloidal MCC, and cream paste MCC, can be obtained depending on the ensuing processing steps utilized during the preparation of MCC (Fig. 3.2).

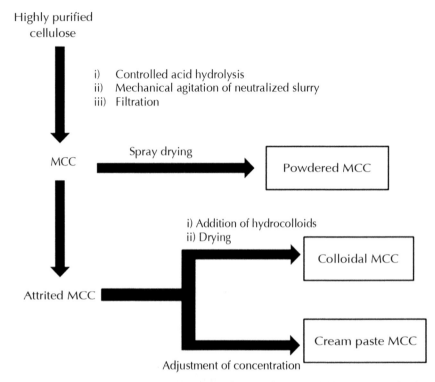

FIGURE 3.2 Schematic diagram describing the manufacture of different types of MCC.

Powdered MCC is obtained from the acid hydrolysis product by subsequent neutralization, filtration, and spray drying. Spray-dried, powdered MCC particles are free flowing and porous in nature as they are aggregates of many primary particles that are held together tightly by hydrogen bonding. Due to their aggregate structure, powdered MCC particles possess many dislocations and slip planes that can fracture and realign during compression. Plastic deformation of the particles during compression enables strong inter-particulate bond formation in a matrix. Capillary action of its pores makes powdered MCC highly absorptive. In addition, its high porosity facilitates wetting and rapid drying.

Colloidal MCC is a co-processed blend of MCC and hydrocolloids. Commonly used hydrocolloids are water-soluble cellulose derivatives or polysaccharides, such as sodium carboxymethyl cellulose and xanthan gum. The hydrocolloids form a coat on the MCC particles which acts as a barrier preventing the agglomeration of MCC particles during the drying step (Fig. 3.2). Being water-soluble, upon addition of water to the colloidal MCC particles, the hydrocolloids hydrate and dissolve allowing independent dispersion of the MCC particles. The dispersed MCC particles form a three-dimensional network structure giving the unique sol-like functionality to these colloidal MCC grades in water. Colloidal MCC suspensions show thixotropic behavior where the hysteresis is more prominent at higher concentrations of MCC.[9] The colloidal MCC has its stability improved through the formation of a three-dimensional network structure in water which prevents syneresis caused by heat and pressure. Different MCC products that are available commercially and used in the pharmaceutical industries are summarized in Tables 3.1–3.2.

3.2.2 INTER-GRADE VARIABILITY IN MCC

As a part of supplier risk management process, it is often necessary to source for an alternative grade of excipients to ensure business continuity. Equivalency between two or more grades of excipients (in this case, MCC) should be demonstrated not only on the physical properties of the excipient but more importantly on its effect related to the drug product manufacturing process control and final product quality. The functional relationship between the excipient properties and the drug product performance is not always well understood to enable the formulator or manufacturing experts to make an informed decision before an alternative grade is procured. More often than not, equivalency is demonstrated from manufacturing pilot scale, if not full scale commercial batches and checking whether the batch manufactured with the new grade meets the end-product specifications. However, this approach may or may not always contribute to better process understanding which becomes critical during root cause analysis or trouble shooting. While all MCC grades for pharmaceutical use fulfill pharmacopeial specifications, their behaviors during the drug product manufacturing are not entirely equivalent. Depending on the source and composition of the plant material (wood pulp) as well as manufacturing conditions such as treatment (e.g., acidic or alkaline treatment, extent of breakdown, and bleaching sequence), methods for mechanical (e.g., spray drying and milling), and chemical (e.g., hydrolysis) processing, the physicochemical properties can differ considerably.

TABLE 3.1 Commercial MCC Products for Pharmaceutical Applications (Powdered MCC).[a,b]

Powdered MCC	Major Brands	Features/Advantages	Main Uses
Standard grade	Avicel® PH 101, Avicel® PH 102	Superior compactibility, self-lubricating, and promote rapid disintegration in tablets	Dry compression and wet granulation. Auxiliary binder in wet granulation, control wet mass consistency
Small particle size grade	Avicel® PH 105	Very small particle size	For tableting of coarse, granular and hard-to-compact materials
Large particle size grade	Avicel® PH 200	Superior flowability, minimum weight variation, excellent content uniformity	Maximum tableting throughput
Low density grade	Ceolus® KG 801, Ceolus® KG 802	High porosity, excellent compactibility, reduces MCC amount required, decreases tablet size, lower compression force needed	Especially good for tableting pressure-sensitive ingredients such as enzymes, antibiotics and film coated granules
High density Grade	Avicel® PH 301, Avicel® PH 302	Increase flow and production rates, reduce capsule weight variation, take up less space in the formulation, result in smaller capsules.	Capsule filling
Low moisture grade	Avicel® PH 103, Avicel® PH 112, Avicel® PH 113	Very low moisture content	Formulation of moisture-sensitive actives

[a]Avicel® PH is a registered trademark of FMC Biopolymer, FMC Corporation.
[b]Ceolus® KG is a registered trademark of Asahi-Kasei Chemicals Corporation, Asahi-Kasei Group.

TABLE 3.2   Commercial MCC Products for Pharmaceutical Applications (Specialized Powdered MCC, MCC Spheres, and Colloidal MCC).[a,b,c]

Specialized MCC Products	Major Brands	Features/Advantages	Main Uses
Specialized powdered MCC MCC + silicon dioxide	Prosolv® 50M, Prosolv® 90M	Excellent flow and compactibility, better content uniformity and enhanced ordered mixing. Require less binder, less glidant and less disintegrant because tablet sizes are smaller.	Prosolv® 50M: wet granulation Prosolv® 90M: dry granulation and capsule filling
Specialized powdered MCC MCC + guar gum	Avicel® CE-15	Superior organoleptic experience, smoother mouth feel, reduced grittiness and tooth packing. Taste masking effects. Tablets produced are softer, less friable and disintegrate rapidly.	Added in 10 % w/w concentration to prepare chewable tablets.
MCC spheres	Celphere® SCP-100, Celphere® CP-203, Celphere® CP-305, Celphere® CP-507	High sphericity and mechanical strength, low reactivity, fine and uniform particle size, maximum water absorption.	Spherical seed cores for coating
Colloidal MCC MCC + sodium carboxy-methyl cellulose	Avicel® RC-591, Avicel® RC 581, Avicel® CL-611	Highly thixotropic, provide suspension with good long term stability against phase separation.	Emulsion and suspension formulation

[a]Prosolv® is a registered trademark of J. Rettenmaier & Söhne Pharma, J. Rettenmaier & Söhne Group.

[b]Celphere® is a registered trademark of Asahi-Kasei Chemicals Corporation, Asahi-Kasei Group.

[c]Avicel® RC is a registered trademark of FMC Biopolymer, FMC Corporation.

Cellulose exists in two polymorphic forms, Cellulose I and Cellulose II. Cellulose I is the native polymorph. Cellulose II is the solubilized and regenerated or recrystallized form of Cellulose I. Compared to conventional MCC-containing Cellulose I, modified MCC grades containing Cellulose II have higher bulk, tapped and true densities, and higher Carr's index and Hausner ratio values.[10] X-ray diffraction studies conducted on various MCC grades obtained from different suppliers showed source-dependent polymorphism of MCC.[11,12] Although all the investigated MCC grades had comparable percent crystallinities, a distinct shoulder around 20.5° (2θ), characteristic of Cellulose II, was observed in the x-ray diffractogram of Unimac MG 100. The remaining MCC grades exhibited peaks at 15, 16 and 22.6° (2θ), characteristic of Cellulose I. In another study, a similar distinct shoulder peak, characteristic of Cellulose II, was observed in the diffractograms of Avicel PH 301, Avicel PH 302, Pharmacel 101, and Pharmacel 102.[6] Besides existence of two polymorphic forms, batch or source variation was also observed in the crystallinity indices of different MCC grades[5] although this difference was not marked in that specific range of MCCs investigated. Landin et al.[13] investigated batch variation from the same supplier by comparing two batches of MCC, differing in pulp type and manufacturing process, to a standard MCC grade. Pulp type and manufacturing process were found to affect the hemicellulose and lignin contents in MCC which influenced the swelling properties (hydrophilic hemicellulose) and hydrophobic behavior of MCC (lignin). The pulp type influenced the specific surface area and pore structure of resultant MCC powders though it did not appear to affect their size and size distribution, compressibility, flow properties, and equilibrium moisture uptake. A marked batch variation in pulp type is envisaged to translate to changes in pore structure and specific surface area of MCC particles and their wetting ability. This in turn affects the process of matrix formation and the quality attributes of the final products.

3.2.3 PHARMACEUTICAL PROCESSES

Multiple research papers have been published on the effect of MCC source variation on tableting performance including tablet tensile strength.[14–16] By and large, different MCC grades yielded tablets with different tensile strengths and compaction profiles. Specifically, the effects of MCC particle size on the tensile strength of tablets were investigated by Khan

Microcrystalline Cellulose: An Overview

and Pilpel[16] but the force-hardness profiles of tablets can also be influenced by other physical factors affecting inter-particulate bonding.

Effects of the type and origin of MCC products on capsule filling were investigated by Patel and Podczeck.[17] A total of eight MCC grades differing mainly in particle size were evaluated and their properties were related to the coefficient of capsule fill weight variation. Ultra-fine MCC grades such as Avicel PH 105 failed to yield properly filled capsules due to extremely poor flowability. Disintegration time of capsules was affected by MCC particle size whereby products of the intermediate size grades (60–70 µm) exhibited noticeably wider variability.

In a study on wet massing (granulation) behaviors, Parker and Rowe[3] attempted to correlate the source of MCC to observed changes in their solid state, thermodynamic and particulate characteristics. Examination of particle morphologies under the scanning electron microscope revealed disparities in particle shape and size of different MCCs. These disparities affect the degree of densification which in turn, governs the liquid requirement during wet massing. Since spherical particles were found to pack better than irregular-shaped particles, the amount of liquid needed for granulation was correspondingly reduced. Comparable enthalpies of immersion suggested that the proportion of anhydroglucose units in amorphous regions of MCC molecules was similar between various MCCs. However, when their surface areas were taken into consideration, differences in their interaction with water became more apparent. For MCCs with a larger surface area, more water molecules were needed to form the multi-molecular layers of free water on their surfaces.[3]

Rheological properties of the wet massed MCCs obtained using mixer torque rheometry were observed to be markedly different for the three MCC grades investigated. These observations concurred with those of Rowe and Sadeghnejad,[1] notwithstanding the smaller number of MCC grades employed in the study. Soh et al.[18] investigated the influence of MCC particle size on MCC rheological properties. A series of six fractionated MCC grades were customized to yield different particle size varieties (36 to 255 µm) of a commercial grade, Comprecel M 101. Despite the marked difference in particle sizes, all the MCC grades required the same amount of water to achieve maximum torque value (W_{Tmax}) in the mean torque against water addition plot. Further, Soh et al.[7] and Liew et al.[19] reported that particle size of MCC did not appear to have an effect on both the W_{Tmax} and the magnitude of maximum torque value (T_{max}) in the mean torque against water addition plot. Similarly, Rowe and Sadeghnejad[1] showed that a change in the particle size of MCC had a minor effect on W_{Tmax} and T_{max}.

From the findings of various researchers, it appears that the source of MCC grades has a significant influence on their rheological properties. Parker and Rowe[3] pointed out that materials of comparable size but from different sources showed distinct differences in W_{Tmax}. However, their study was met with limitation as only three different MCC grades, two commercial grades and one manufactured from bacterial source, were used. Chohan and Newton compared rheological properties of MCC grades from different sources but with similar particle size and established that physicochemical characteristics based on the source of cellulose other than particle size had a significant effect on their rheological properties.[20–21] Soh et al.[7] categorized 11 MCC grades with different physical properties from five different suppliers into different groups on the basis of their interaction with water. It was observed that two high density MCC grades, Avicel PH 301 and Avicel PH 302, exhibited significantly lower W_{Tmax} and higher T_{max} values. However, the remaining nine MCC grades were shown to have similar W_{Tmax} and T_{max} values.

3.2.4 MCC AS PELLETIZATION AID

A material to be used as a pelletization aid must possess attributes such as insolubility in the moistening liquid, high liquid absorption and retention capacities, good binding properties, and sufficiently large surface area.[22] It was postulated that the cohesion of the cellulose microcrystals is largely due to the van der Waals attraction between hydrogen-bonded sheets.[23,24] Although van der Waals attraction between hydrogen-bonded sheets may contribute to cellulose crystal cohesion, the hydroxyl groups in cellulose and their ability to hydrogen bond are deemed to play a major role not only in directing the crystalline packing but also in governing the different physical characteristics of cellulose.[25,26]

3.3.1 MODELS EXPLAINING PELLETIZATION AID FUNCTIONALITY OF MCC

MCC is able to absorb and retain a large quantity of liquid due to its large surface area and high internal porosity.[27] In addition, it has good binding properties and high cohesiveness. All these properties contribute to the behavior of MCC as an excellent pelletization aid. So far, two models have been reported to explain the excellent pelletization aid functionality of MCC.

Fielden et al.[28] described MCC as a molecular sponge owing to its ability to store a large amount of liquid, especially water. The sponge-like activity allows this stored water to be released during evaporation. This sponge model has its basis in the various stages of liquid saturation. This feature is believed to be crucial in controlling water movement or distribution within wetted powder masses that are to be subsequently extruded and spheronized. During extrusion, the sponges are compressed and the stored water is squeezed out to lubricate the material flowing through the extruder screen.

The crystallite gel model, an alternative to the sponge model, was presented by Kleinebudde.[29] The model proposed that MCC particles in the formulation are broken up into smaller particles by shear forces acting on the particles during extrusion in the presence of a moistening liquid, particularly water. Single crystallites of colloidal size are introduced with increasing shear stress. It was hypothesized that these colloidal particles are able to form a delicate crystallite gel network by cross-linking via hydrogen bonds at the amorphous end and immobilizing the liquid.

3.3.2 PELLETIZATION WITH MCC USING WATER AS MOISTENING LIQUID

Investigations on the influence of MCC particle size on the quality of pellets produced by the extrusion–spheronization process found that pellet sphericity, yield, size, and size distribution were independent of the particle size of MCC grades used.[30] Furthermore, surface roughness of pellets produced from formulations containing different MCC grades was not observed to be dependent on MCC particle size.[31]

The various MCC grades with different particle sizes are aggregates constituting different numbers of small sub-units that are held together by hydrogen bonds.[32] The small sub-units are released from the aggregates during sonication treatment of their suspensions in water[32] and during wet processing.[33,34] It was observed that the different MCC grades, irrespective of their different dry state particle sizes, de-aggregate during the different wet processing steps of extrusion–spheronization, giving rise to in-process particles with more or less similar sizes (12–22 μm).[34] Having more or less similar in-process particle sizes, formulations containing the different MCC grades exhibited similar rheological properties and produced extrudates with similar strength and plasticity. During the spheronization step, these extrudates subsequently yielded pellets with comparable quality in terms of sphericity, yield, size, and size distribution.

Heng and Koo[6] characterized the physical properties of a more extensive range of MCC grades from different suppliers and correlated their physical properties with their performance in extrusion–spheronization. Results showed that particle size, size distribution, porosity, and source were found to have little influence on the produced pellet quality. However, some critical properties, namely packing densities and pore volumes, were identified as having strong influences on water requirement for successful pelletization and produced pellet quality. Grades with good packing densities required comparatively lower amounts of moistening liquid to form coherent extrudates as well as pellets of equivalent size compared to those with lower packing densities.

The pelletization aid functionality of MCC containing Cellulose II grades has been investigated.[35–38] It was observed that the amount of Cellulose II required for the production of reasonably good quality pellets was comparable to that of Cellulose I.[38] However, compared to MCC with Cellulose I polymorph, MCC with Cellulose II polymorph required less amount of water for the formation of spherical pellets. Lower water requirement for MCC with Cellulose II polymorph in pelletization was postulated to be related to its lower water-binding capacity. Spheronization mechanism of MCC with Cellulose II formulation was investigated and compared to that with Cellulose I.[37] A different spheronization mechanism with Cellulose II formulation was noted. In the case of the formulation containing Cellulose II, pellet weight and equivalent diameter decreased slightly at the initial stage of spheronization and thereafter increased with progression of the spheronization process. The opposite trend was observed for the amount of the fines fraction. In the case of the formulation containing MCC with Cellulose I, pellet weight remained almost constant but equivalent diameter decreased with spheronization time. Based on the pellet characteristics at different time points during spheronization, a new spheronization mechanism was proposed for

Cellulose II. It was postulated that small pellets abraded and layered on the other pellets during spheronization. The resultant pellets had higher porosity and lower tensile strength, and exhibited faster disintegration and drug release properties than those produced from MCC grades containing Cellulose I.[38] With increasing MCC (Cellulose I or II) proportion in the formulation, a larger amount of water was required for spheronization and pellets produced had larger diameter, lower porosity, and longer disintegration time[30,36] In general, MCC-based pellets formed by extrusion–spheronization using water as the moistening liquid exhibit high yield, low span, high sphericity, and high mechanical strength[39] and do not swell or disintegrate readily when they are placed in an aqueous solution.[40,41] *In vitro* dissolution tests showed that these MCC-based pellets exhibited sustained drug release properties.[42,43]

3.3.3 PELLETIZATION WITH MCC USING NON-AQUEOUS SOLVENTS AS MOISTENING LIQUID

Besides water, ethyl alcohol, isopropyl alcohol (IPA), and dimethyl sulfoxide have also been used as the moistening liquid in investigations on pellet production by extrusion–spheronization. Millili and Schwartz investigated the utility of various ethyl alcohol–water solvent mixtures as moistening liquid for the production of pellets by extrusion–spheronization using a powder blend containing 90% MCC.[44] Pellets could be formed when moistening liquid contained ethyl alcohol up to 95%, but not when absolute ethyl alcohol was used. Furthermore, it was observed that pellet hardness and sphericity increased as the concentration of ethyl alcohol decreased in the ethyl alcohol–water mixture. It was suggested that the differences in hardness based on the moistening liquid composition might be due to changes in the extent of both intra- and inter-molecular hydrogen bonding within the hydroxyl groups of the MCC.[45–47] In contrast, Millili et al. found no significant difference in the x-ray diffraction patterns, degree of crystallinity or internal energies of oven-dried MCC-based pellets prepared using an ethyl alcohol–water solvent mixture, deuterated water, or water as moistening liquid.[48]

Chatlapalli and Rohera investigated the applicability of IPA as moistening liquid for the production of MCC-based pellets.[49] When IPA was used as a moistening liquid, the pellets were highly friable and tended to crumble into powder during drying and subsequent handling.[49] This problem was overcome by the addition of a binder, hydroxypropyl cellulose (5% w/w) which is soluble in IPA.

Schroder and Kleinebudde found that the structure of MCC-based pellets was markedly influenced by the concentrations of IPA in the IPA–water mixture used as moistening liquid.[50] In the investigated IPA concentration range (0–60%), the porosity of pellets increased while the crushing strength decreased with increasing concentrations of IPA up to 40%. A kink was observed in the plot of pellet porosity or crushing strength against IPA concentration at around 40% IPA concentration. Furthermore, IPA concentrations exceeding 40% in the moistening liquid resulted in fast and complete disintegration of the resultant pellets compared to those prepared with lower fractions of IPA in the solvent mixture. This fast disintegration was accompanied by rapid drug dissolution from pellets. It was postulated that this phenomenon was due to the change in the MCC particle bonding at concentration of 40% IPA.

As MCC particle de-aggregation involves hydrogen bond breaking, moistening liquids with lower polarity, such as water–alcohol mixtures with higher alcohol proportions, induced lesser de-aggregation and yielded MCC with larger particle sizes.[51] When such water–alcohol mixtures were employed during extrusion–spheronization with MCC, the larger particle size of MCC and lower surface tension of the moistening liquid gave rise to moistened masses with lower cohesive strength. During pelletization, agglomerate growth by coalescence and closer packing of components by particle rearrangement would be limited. Thus, weaker, less spherical pellets with smaller size and wider size distribution were produced.

Mascia et al. introduced dimethyl sulfoxide as a non-aqueous moistening liquid for pelletization by extrusion–spheronization.[52] Dimethyl sulfoxide is a polar solvent and it is used in a number of finished pharmaceutical products. However, it is not currently permitted for use in oral solid dosage forms, although it is a class III solvent (International Conference on Harmonisation, ICH, Guidelines) along with ethyl alcohol and acetone. In the study by Mascia et al.,[52] the functionality of dimethyl sulfoxide as moistening liquid was compared with water and ethyl alcohol for MCC-based pellet formation. MCC-moistened mass, prepared using dimethyl sulfoxide as moistening liquid, was successfully processed to produce highly spherical pellets. The mechanical strength of pellets prepared using dimethyl sulfoxide was comparable to those prepared using water. The authors pointed out that polarity, surface tension, and viscosity of the moistening liquid were the critical properties of the solvent to be used as moistening liquid for successful pelletization by extrusion–spheronization. On the basis of the similar microstructure of pellets formulated using dimethyl sulfoxide and water as moistening liquids, it was postulated that dimethyl sulfoxide interacts with MCC in a similar fashion to water. Ethyl alcohol, on the other hand, could not solvate the MCC polar group unlike dimethyl sulfoxide and water as a result of its low polarity and surface tension attributes. Ethyl alcohol was retained in the large inter-particular void spaces and was not penetrated into the small micropores in MCC. Ethyl alcohol could not form hydrogen bonds with the microfibrils and did not promote microcrystal rearrangement. The microstructure of pellets formed was different from those of water and dimethyl sulfoxide.

3.3.4 MCC–SOLVENT INTERACTION

Several investigations were carried out to address the moistening liquid selectivity of MCC-based formulations for successful production of pellets. Chernoberezhskii et al. studied the aggregation stability of MCC aqueous dispersions over a wide range of pH (1–11) using flow ultra-microscopy.[53] The molecular and ion-electrostatic components of the inter-particle interaction energy were calculated and were compared with those obtained experimentally. A discrepancy between the calculated and the experimental results of the energy of inter-particle interaction was observed. It was assumed that this discrepancy could be due to additional attractive forces between the MCC particles which might be attributed to the hydrogen bonding or the dipole–dipole interactions between hydrated MCC particles.

Attenuated total reflection infrared spectroscopy was also used to investigate the interactions between MCC and water.[54] Based on the results, total drying time was divided into four stages. Evaporation of bulk water predominated during the first two stages, i.e., 0–11 and 14–41 min, whereas loss of absorbed water in MCC occurred during the last two stages, i.e., 49–80 and 101–195 min. The disruption of inter-chain hydrogen bonds, intra-chain hydrogen bonds, and hydrogen bonds in Iβ and Iα crystal forms was particularly observed in the last stage of drying. These results indicated that the water molecules and hydroxyl groups of

MCC constructed hydrogen bonds that stabilized the network of MCC. The structural feature of MCC was changed in the presence of water as well as after removal of absorbed water.

To understand the effect of granulation on the physical state of water in MCC, thermo-porosimetry, along with the solute exclusion technique, was used for the measurement of different water fractions and pore size distributions of both ungranulated and granulated wet masses.[55] The granulated MCC wet masses were prepared in a planetary mixer, whereas ungranulated MCC wet masses were made in a mortar. In MCC wet masses, four fractions of water, i.e., bulk, free, freezing bound, and non-freezing water, were noticed. Granulation increased the volume of free water and freezing bound water, whereas it decreased the volume of bulk water. On the basis of the obtained results, it was assumed that granulation might help to increase the volume of micropores in MCC and convert them into macropores to contain free water.

3.3 FUTURE PERSPECTIVES

Oral colonic drug delivery is a popular mode of drug administration with high tissue specificity. It is recognized to be therapeutically advantageous for the treatment of ulcerative colitis, Crohn's disease, and colon cancer.[56] MCC-based pellets have been designed for the said purpose with a good relationship between in vitro drug release and in vivo pharmacokinetics and pharmacodynamics.[57,58] A more comprehensive research in relation to MCC physical properties is essential to further improve the drug-targeting attribute of MCC-based pellets.

KEYWORDS

- extrusion
- microcrystalline cellulose
- pelletization
- spheronization

REFERENCES

1. Rowe, R. C.; Sadeghnejad, G. R. The Rheology of Microcrystalline Cellulose Powder/Water Mixes–Measurement Using a Mixer Torque Rheometer. *Int. J. Pharm.* **1987**, *38*, 227–229.
2. Staniforth, J. N.; Baichwal, A. R.; Hart, J. P.; Heng, P. W. S. Effect of Addition of Water on the Rheological and Mechanical Properties of Microcrystalline Celluloses. *Int. J. Pharm.* **1988**, *41*, 231–236.
3. Parker, M. D.; Rowe, R. C. Source Variation in the Wet Massing (granulation) of Some Microcrystalline Celluloses. *Powder Technol.* **1991**, *65*, 273–281.
4. Parker, M. D.; York, P.; Rowe, R. C. Binder-Substrate Interactions in Wet Granulation. 3: The Effect of Excipient Source Variation. *Int. J. Pharm.* **1992**, *80*, 179–190.
5. Rowe, R. C.; McKillop, A. G.; Bray, D. The Effect of Batch and Source Variation on the Crystallinity of Microcrystalline Cellulose. *Int. J. Pharm.* **1994**, *101*, 169–172.
6. Heng, P. W. S.; Koo, O. M. Y. A Study of the Effects of the Physical Characteristics of Microcrystalline Cellulose on Performance in Extrusion Spheronization. *Pharm. Res.* **2001**, *18*, 480–487.
7. Soh, J. L. P.; Chen, F.; Liew, C. V.; Shi, D.; Heng, P. W. S. A Novel Preformulation Tool to Group Microcrystalline Celluloses Using Artificial Neural Network and Data Clustering. *Pharm. Res.* **2004**, *21*, 2360–2368.
8. Iijima, H.; Takeo, K. Microcrystalline Cellulose. In *Handbook of Hydrocolloids;* Phillips, G., Williams, P., Eds.; CRC Press: Washington, 2000; pp 331–346.
9. Falkiewicz, M. J. Rheology. Fundamental Principles in Product Development. *Soap Cosmetics Chem. Spec.* **1980**, *56*, 46–70.
10. Kumar, V.; de la Luz Reus-Medina, M.; Yang, D. Preparation, Characterization, and Tabletting Properties of a New Cellulose-Based Pharmaceutical Aid. *Int. J. Pharm.* **2002**, *235*, 129–140.
11. Chatrath, M.; Staniforth, J. N.; Herbert, I.; Luk, S. Y.; Richards, G. Source-Dependent Polymorphism of Microcrystalline Cellulose. *J. Pharm. Pharmacol. Suppl.* **1991**, *43*, 4.
12. Landin, M. et al. Effect of Country of Origin on the Properties of Microcrystalline Cellulose. *Int. J. Pharm.* **1993**, *91*, 123–131.
13. Landin, M. et al. Effect of Batch Variation and Source of Pulp on the Properties of Microcrystalline Cellulose. *Int. J. Pharm.* **1993**, *91*, 133–141.
14. Doelker, E.; Gurny, R.; Schurz, J.; Jánosi, A.; Matin, N. Degrees of Crystallinity and Polymerization of Modified Cellulose Powders for Direct Tableting. *Powder Technol.* **1987**, *52*, 207–213.
15. Whiteman, M.; Yarwood, R. J. Variations in the Properties of Microcrystalline Cellulose from Different Sources. *Powder Technol.* **1988**, *54*, 71–74.
16. Khan, F.; Pilpel, N. The Effect of Particle Size and Moisture on the Tensile Strength of Microcrystalline Cellulose Powder. *Powder Technol.* **1986**, *48*, 145–150.
17. Patel, R.; Podczeck, F. Investigation of the Effect of Type and Source of Microcrystalline Cellulose on Capsule Filling. *Int. J. Pharm.* **1996**, *128*, 123–127.
18. Soh, J. L. P.; Yang, L.; Liew, C. V.; Cui, F. D.; Heng, P. W. S. Importance of Small Pores in Microcrystalline Cellulose for Controlling Water Distribution During Extrusion - Spheronization. *AAPS Pharm. Sci. Tech.* **2008**, *9*, 972–981.

19. Liew, C. V.; Soh, J. L. P.; Chen, F.; Shi, D.; Heng, P. W. S. Application of Multidimensional Scaling to Preformulation Sciences: A Discriminatory Tool to Group Microcrystalline Celluloses. *Chem. Pharm. Bull.* **2005**, *53*, 1227–1231.
20. Chohan, R. K.; Newton, J. M. Analysis of Extrusion of Some Wet Powder Masses Used in Extrusion/Spheronisation. *Int. J. Pharm.* **1996**, *131*, 201–207.
21. Newton, J. M.; Chow, A. K.; Jeewa, K. B. The Effect of Excipient Source on Spherical Granules Made by Extrusion/Spheronization. *Pharm. Technol. Int.* **1992**, *4*, 52–58.
22. Liew, C. V.; Gu, L.; Soh, J. L. P.; Heng, P. W. S. Functionality of Cross-Linked Polyvinylpyrrolidone as a Spheronization Aid: A Promising Alternative to Microcrystalline Cellulose. *Pharm. Res.* **2005**, *22*, 1387–1388.
23. French, A. D.; Miller, D. P.; Aabloo, A. Miniature Crystal Models of Cellulose Polymorphs and other Carbohydrates. *Int. J. Biol. Macromolec.* **1993**, *15*, 30–36.
24. Cousins, S. K.; Brown Jr, R. M. Cellulose I Microfibril Assembly: Computational Molecular Mechanics Energy Analysis Favours Bonding by Van Der Waals Forces as the Initial Step in Crystallization. *Polymer.* **1995**, *36*, 3885–3888.
25. Nishiyama, Y.; Sugiyama, J.; Chanzy, H.; Langan, P. Crystal Structure and Hydrogen Bonding System in Cellulose Iα from Synchrotron X-ray and Neutron Fiber Diffraction. *J. Am. Chem. Soc.* **2003**, *125*, 14300–14306.
26. John, M. J.; Thomas, S. Biofibres and Biocomposites. *Carbohydr. Polym.* **2008**, *71*, 343–364.
27. Shah, R. D.; Kabadi, M.; Pope, D. G.; Augsburger, L. L. Physico-Mechanical Characterization of the Extrusion-Spheronization Process. Part II: Rheological Determinants for Successful Extrusion and Spheronization. *Pharm. Res.* **1995**, *12*, 496–507.
28. Fielden, K. E.; Newton, J. M.; O'Brien, P.; Rowe, R. C. Thermal Studies on the Interaction of Water and Microcrystalline Cellulose. *J. Pharm. Pharmacol.* **1988**, *40*, 674–678.
29. Kleinebudde, P. The Crystallite-Gel-Model for Microcrystalline Cellulose in Wet-Granulation, Extrusion, and Spheronization. *Pharm. Res.* **1997**, *14*, 804–809.
30. Sinha, V. R.; Agrawal, M. K.; Kumria, R. Influence of Formulation and Excipient Variables on the Pellet Properties Prepared by Extrusion Spheronization. *Curr. Drug Deliv.* **2005**, *2*, 1–8.
31. Sarkar, S.; Ang, B. H.; Liew, C. V. Influence of Starting Material Particle Size on Pellet Surface Roughness. *AAPS Pharm. Sci. Tech.* **2014**, *15*, 131–139.
32. Ek, R.; Alderborn, G.; Nyström, C. Particle Analysis of Microcrystalline Cellulose: Differentiation between Individual Particles and their Agglomerates. *Int. J. Pharm.* **1994**, *111*, 43–50.
33. Brittain, H. G.; Lewen, G.; Newman, A. W.; Fiorelli, K.; Bogdanowich, S. Changes in Material Properties Accompanying the National Formulary (NF) Identity Test for Microcrystalline Cellulose. *Pharma. Res.* **1993**, *10*, 61–67.
34. Sarkar, S.; Heng, P. W. S.; Liew, C. V. Insights into the Functionality of Pelletization Aid in Pelletization by Extrusion-Spheronization. *Pharm. Dev. Technol.* **2013**, *18*, 61–72.
35. Rojas, J.; Kumar, V. Evaluation of Microcrystalline Cellulose II (MCCII) as an Alternative Extrusion-Spheronization Aid. *Pharmazie.* **2012**, *67*, 595–597.
36. Krueger, C.; Thommes, M.; Kleinebudde, P. Influence of MCC II Fraction and Storage Conditions on Pellet Properties. *Eur. J. Pharm. Biopharm.* **2013**, *85*, 1039–1045.

37. Krueger, C.; Kleinebudde, M. T. P.; Krueger, C.; Thommes, M.; Kleinebudde, P. Spheronisation Mechanism of MCC II-Based Pellets. *Powder Technol.* **2012**, *238*, 176–187.
38. Krueger, C.; Kleinebudde, M. T. P. 'MCC SANAQ Burst''–A New Type of Cellulose and its Suitability to Prepare Fast Disintegrating Pellets. *J. Pharm. Innov.* **2010**, *5*, 45–57.
39. Rowe, R. C. Spheronization–A Novel Pill-Making Process. *Pharm. Int.* **1985**, *6*, 119–123.
40. Okada, S.; Nakahara, H.; Isaka, H. Adsorption of Drugs on Microcrystalline Cellulose Suspended in Aqueous Solutions. *Chem. Pharm. Bull.* **1987**, *35*, 761–768.
41. Schröder, M. K. P. Development of Disintegrating Pellets Obtained from Extrusion/Spheronization. *J. Pharm. Sci.* **1995**, *1*, 415–418.
42. O'Connor, R. E.; Schwartz, J. B. Spheronization II: Drug Release from Drug-Diluent Mixtures. *Drug Dev. Ind. Pharm.* **1985**, *11*, 1837–1857.
43. Zimm, K. R.; Schwartz, J. B.; O'Connor, R. E. Drug Release from a Multiparticulate Pellet System. *Pharm. Dev. Technol.* **1996**, *1*, 37–42.
44. Millili, G. P.; Schwartz, J. B. The Strength of Microcrystalline Cellulose Pellets: The Effect of Granulating with Water/Ethanol Mixtures. *Drug Dev. Ind. Pharm.* **1990**, *16*, 1411–1426.
45. Nakai, Y.; Fukuoka, E.; Nakajima, S.; Yamamoto, K. Crystallinity and Physical Characteristics of Microcrystalline Cellulose. II. Fine Structure of Ground Microcrystalline Cellulose. *Chem. Pharm. Bull.* **1977**, *25*, 2490–2496.
46. Huttenrauch, R. Identification of Hydrogen Bonds in Drug Forms by Means of Deuterium Exchange Demonstration of Binding Forces in Compressed Cellulose Forms. *Pharmazie.* **1971**, *26*, 645–646.
47. Reier, G. E.; Shangraw, R. Microcrystalline Cellulose in Tableting. *J. Pharm. Sci.* **1966**, *55*, 510–514.
48. Millili, G. P.; Wigent, R. J.; Schwartz, J. B. Differences in the Mechanical Strength of Dried Microcrystalline Cellulose Pellets are not due to Significant Changes in the Degree of Hydrogen Bonding. *Pharm. Dev. Technol.* **1996**, *1*, 239–249.
49. Chatlapalli, R.; Rohera, B. D. Physical Characterization of HPMC and HEC and Investigation of their Use as Pelletization Aids. *Int. J. Pharm.* **1998**, *161*, 179–193.
50. Schroder, M.; Kleinebudde, P. Structure of Disintegrating Pellets with Regard to Fractal Geometry. *Pharm. Res.* **1995**, *12*, 1694–1700.
51. Sarkar, S.; Liew, C. V. Moistening Liquid-Dependent De-Aggregation of Microcrystalline Cellulose and its Impact on Pellet Formation by Extrusion-Spheronization. *AAPS Pharm. Sci. Tech.* **2014**, *15*, 753–761.
52. Mascia, S.; Seiler, C.; Fitzpatrick, S.; Wilson, D. I. Extrusion-Spheronisation of Microcrystalline Cellulose Pastes Using a Non-Aqueous Liquid Binder. *Int. J. Pharm.* **2010**, *389*, 1–9.
53. Chernoberezhskii, Y. M.; Baturenko, D. Y.; Lorentsson, A. V.; Zhukov, A. N. Aggregation Stability of Aqueous Dispersions of Microcrystalline Cellulose: Ph Dependence. *Colloid J. Rus. Acad. Sci.: Kolloidnyi Zhurnal.* **2003**, *65*, 390–393.
54. Watanabe, A. et al. Drying Process of Microcrystalline Cellulose Studied by Attenuated Total Reflection IR Spectroscopy with Two-Dimensional Correlation Spectroscopy and Principal Component Analysis. *J. Molec. Struct.* **2006**, *799*, 102–110.

55. Luukkonen, P.; Maloney, T.; Rantanen, J.; Paulapuro, H.; Yliruusi, J. Microcrystalline Cellulose-Water Interaction-a Novel Approach Using Thermoporosimetry. *Pharm. Res.* **2001,** *18,* 1562–1569.
56. Wong, T. W.; Colombo, G.; Sonvico, F. Pectin Matrix as Oral Drug Delivery Vehicle for Colon Cancer Treatment. *AAPS Pharm. Sci. Tech.* **2011,** *12* (1), 201–214.
57. Elyagoby, A.; Layas, N.; Wong, T. W. Colon-Specific Delivery of 5-Fluorouracil from Zinc Pectinate Pellets Through *In Situ* Intra-Capsular Ethylcellulose-Pectin Plug Formation. *J. Pharm. Sci.* **2013,** *102* (2), 604–616.
58. Bose, A.; Elyagoby, A.; Wong, T. W. Oral 5-Fluorouracil Colon-Specific Delivery Through In Vivo Pellet Coating for Colon Cancer and Aberrant Crypt Foci Treatment. Int. J. Pharm. **2014,** 468, 178–186.

CHAPTER 4

BIO-BASED PHENOL FORMALDEHYDE FROM LIGNOCELLULOSIC BIOMASS

RASIDI ROSLAN[1], SARANI ZAKARIA[2,*], CHIN HUA CHIA[2], UMAR ADLI AMRAN[2], and SHARIFAH NABIHAH SYED JAAFAR[2]

[1]Faculty of Industrial Sciences and Technology, Universiti Malaysia Pahang, Lebuhraya Tun Razak, Gambang Kuantan 26300, Pahang, Malaysia

[2]Bioresouces and Biorefinery Laboratory, Faculty of Science and Technology, Universiti Kebangsaan Malaysia, Bangi 43600, Selangor, Malaysia

[*]Corresponding author. E-mail: szakaria@ukm.edu.my;, sarani_ zakaria@yahoo.com

CONTENTS

Abstract 76
4.1   Introduction 76
4.2   Production of Bio-Based Phenolic Resin 77
4.3   Type of Resin and Its Chemistry 77
4.4   Biomass Feedstock for the Production of Phenolic Resin 83
4.5   Production of Phenolic Compounds from Biomass 86
4.6   Future Perspectives 92
Keywords 93
References 93

ABSTRACT

Utilization of lignocellulosic biomass in the production of bio-based phenol formaldehyde (PF) resin has received a lot of attention over the last decades as a promising raw material to substitute petroleum-based phenol. This chapter provides a general overview of the production of phenolic compound from biomass via several processes such as fast pyrolysis, vacuum pyrolysis, and liquefaction. Various feedstocks from hardwood, softwood, bark, and agricultural waste have been tested for their suitability as phenolics sources. The PF chemistry, biomass structure, and detail methodology are discussed. Their applications and mechanical properties, including internal bonding, modulus of elasticity, and modulus of rapture and flexural testing of the PF resin are presented. Considerable progress has been reported on the utilization of biomass in the production of PF resins. However, still a lot of effort is required to make the bio-based PF resin competitive with the commercial PF.

4.1 INTRODUCTION

Thermosetting resins play a vital role in current industry due to their high flexibility to be designed into various desired properties, including high modulus, strength, durability, thermal resistance, and chemical resistance, and due to their high crosslinking density.[1] Phenol formaldehyde (PF), which is also called phenolic, still retains its commercial and industrial interest despite the emergence of new thermosets and high-performance polymers.[2] PF resins are widely used as adhesive in the production of wood panel products and many other engineering materials.[3]

PF resin is produced via polycondensation reaction of phenol and form-aldehyde and it is the first synthetic polymer developed commercially, in 1908.[4] PF can be divided into two types: novolac and resol. Chemical properties of PF resin depend on several factors, such as formaldehyde to phenol molar ratio and catalyst used during the polymerization process. Acidic catalyst will produce thermoplastic PF resin, namely novolac, while resol type PF resin can be produced by the reaction of phenol and formaldehyde in the presence of alkaline catalysts. Resol PF is classified as thermosetting resin because it is cured once exposed to heat.[5] There are three main differences between the reaction of formaldehyde with phenol in acidic and alkaline catalysts. The first one is the rate of aldehyde attacks on phenol; the second is the subsequent condensation of phenolic alcohols; and the third is the nature of the condensation reaction.[4]

4.2 PRODUCTION OF BIO-BASED PHENOLIC RESIN

Lignocellulosic biomass is the world's most abundant and promising raw material to substitute petroleum-based chemicals. It comprises three main chemical components: cellulose, hemicelluloses, and lignin.[5] Over the past decades, several attempts were introduced to utilize lignocellulosic biomass, such as wood,[6] bark,[7] lignin, and tannins,[8] to produce PF resins in order to reduce the consumption of petroleum-based phenol.

Hassan et al. studied the resol PF resins produced from the liquefaction of wood in phenol and the resin was applied on bond particle board.[9] The liquefied wood product was reacted with formaldehyde, sodium hydroxide (NaOH), and water, and the reaction was carried out at 60 °C for 5 h. The properties of the wood panels produced using the PF resin are comparable with urea formaldehyde (UF) resin in terms of fracture and elastic modulus. However, the internal bond strength of the panel is lower than that of UF resin, which is mainly due to incomplete curing of the PF resin in the core layer. Longer curing time was suggested by extending the hot pressing process to allow the resin to be fully maintained in the core.

Catalysts that are used to synthesize resol PF resins have also significantly influenced the molecular structure of the resin. Yuan et al. studied the effect of alkali catalysts on the properties of resol resins synthesized from the liquefaction of larch bark.[7] Resol PF resins were produced by 2–3 steps addition of NaOH. The shear strength in wet-dry-wet conditions and free formaldehyde release were investigated. They found that shear strength and wood failure decreased after the third step of NaOH addition. Free formaldehyde emission was higher than the control PF but still fulfilled the industrial standards of China and Japan.

4.3 TYPE OF RESIN AND ITS CHEMISTRY

Understanding on the reaction of phenol with formaldehyde via acidic or basic catalyst is essential to determine the type of PF resin produced. The chemistry of the polymerization process of PF resin is governed by many factors including:

i.   formaldehyde to phenol molar ratio.
ii. liquids, solids, and dispersion.
iii. mode of catalysis: acid, alkali, salt, metal, or enzyme.

The details on the parameters of the preparation of various PF resins are presented in Table 4.1. The table presents the variation for the production of phenolic resin with different types of resins, catalysts, formaldehyde to phenol ratio, physical state of the resin, functional groups, curing method, and resin stability. The presence of acidic and alkaline catalysts during the polymerization of phenol and formaldehyde helps in the production of novalac and resol PF resin, respectively. Resol is classified as thermosetting resin after curing at high temperature. Apart from the use of acidic and basic catalysts, transition metal salt catalyst under mild acidic condition has proven ability to produce high-ortho novolac PF resin.

4.3.1   NOVOLAC PF

Novolac resin can be produced by reacting phenol and formaldehyde at a molar ratio of formaldehyde to phenol <1 using acid catalyst. Molecules of novolac resin do not contain methylol reactive groups and without hardening agent there is no reaction between novolac molecules at high temperature. To complete the resinification, further addition of formaldehyde is required to enable the crosslinking process.

Phenolic ring is less reactive as a nucleophilic center at acidic condition because of the presence of hydroxyl in which protonation tends to occur at the phenol ring as shown in Figure 4.1. However, in Figure 4.2, it is shown that the low reactivity can be overcome as the aldehyde will be

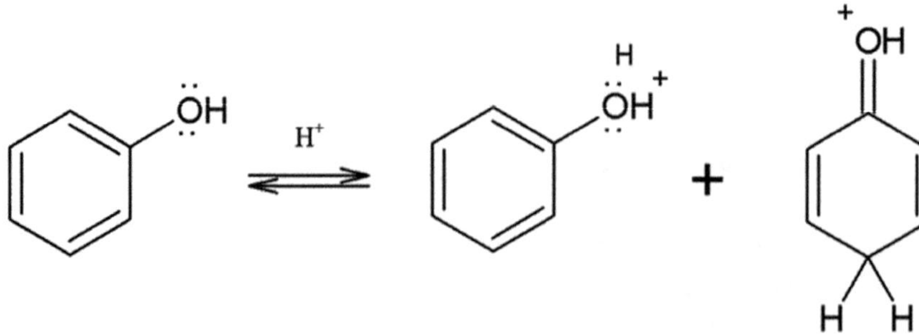

FIGURE 4.1   Protonation of phenol ring in the presence of acidic catalyst.

TABLE 4.1 Phenolic Resin From Different Catalysts.[10]

		Functionality f_{actual}						
Catalyst	Resin Type	F/P	F	P	Physical State	Product Stability	Functional Groups	Mode of Cure
Base	Resol	≥1	<2	≥3	Liquid, solid, solution	Limited	Methylol, phenolic	Acid, base, thermal
Acid	Novolac	<1	2	1.49–1.72	Solid	Stable	Phenolic	Hexa (CH_2O)
Metal salt	Resol or novolac	≥1	High	Ortho	Liquid, solid	Liquid-limited, solid-stable	methylol, phenolic	Same as resol or novolac
Enzyme	Pseudo novolac	–	–	–	Solid	Stable	Phenolic	Resin transformation

protonated in the presence of acid catalyst. Protonated aldehyde will form more effective electrophile for the next reaction to be taken place. Substitution reaction proceeds slowly and the process of condensation occurs as a result of further protonation. The benzyl-carbonium ion is also formed which serves as a nucleophile as shown in Figure 4.3.

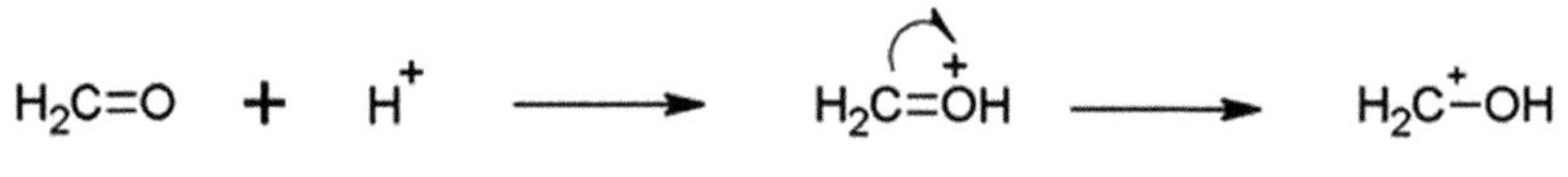

FIGURE 4.2 Protonation of aldehyde in acidic condition.

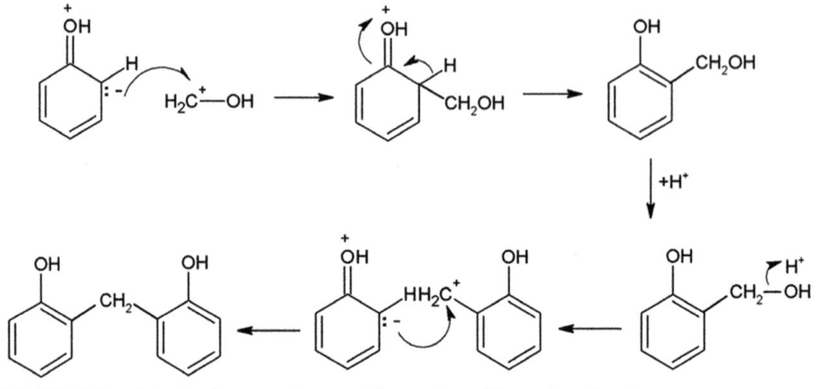

FIGURE 4.3 Substitution reaction and formation of benzyl-carbonium.

4.3.2 RESOL PF

Resol-type PF resin is prepared via equivalent molar or higher ratio of formaldehyde to phenol in the presence of alkaline catalyst. When phenol reacts with formaldehyde under this condition, the hydroxyl groups will be deprotonated into reactive phenoxide ion which is resonance stabilized (Fig. 4.4). The electron density in the phenoxide ion results in an electrophilic aromatic substitution, where there are possibilities to obtain ortho or para methylol phenol. One molecule of resol contains one reactive methylol group. Heating will cause the resol reactive molecule to be condensed to form a larger molecule without the addition of a hardener. This is the basic phenolic component that will be cured into a PF resin.[10]

Condensation of hydroxymethyl moiety is known as step growth polymerization reaction which produces water as a byproduct. According

Bio-Based Phenol Formaldehyde from Lignocellulosic Biomass 81

to Jones, quinone methide is presented as an intermediate during the condensation reaction.[11] Quinone methide is very reactive and readily reacts with nucleophilic sites of either phenol (I) or methylol phenol (II) via electrophilic substitution to form methylene bridge (Fig. 4.4).

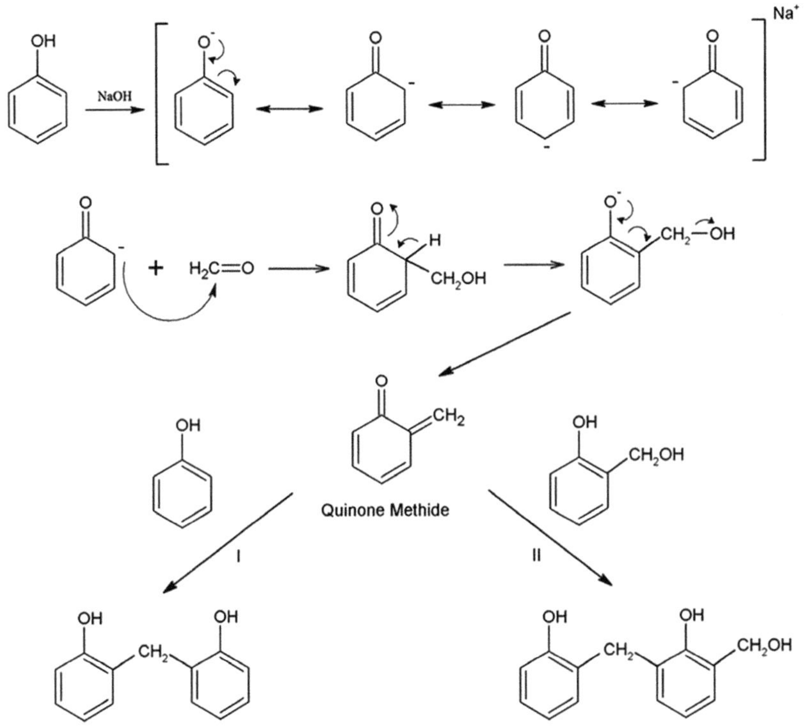

FIGURE 4.4 Addition reaction of formaldehyde to ortho methylol phenol.

Besides methylol phenols, it is also possible to obtain dimethylol phenol and trimethylol phenol (Fig. 4.5), which only occur when the reaction of phenol is carried out at the ratio of formaldehyde to phenol (F/P) $\leq$ 1. These mono-, di-, and tri-hydroxymethyl phenols (HMP) act as temporary intermediates before experiencing oligomerization into dimeric, trimeric, and higher oligomers, depending on the amount of formaldehyde, pH, and temperature of the reaction. These types of intermediates have been examined by Higuichi and co-workers in a series of papers, in which the kinetics fate of 2-HMP, 4-HMP, 2,4-dihydroxymethyl phenol (DHMP), and 2,4,6-trihydroxymethyl phenol (THMP) were all discussed.[12–17]

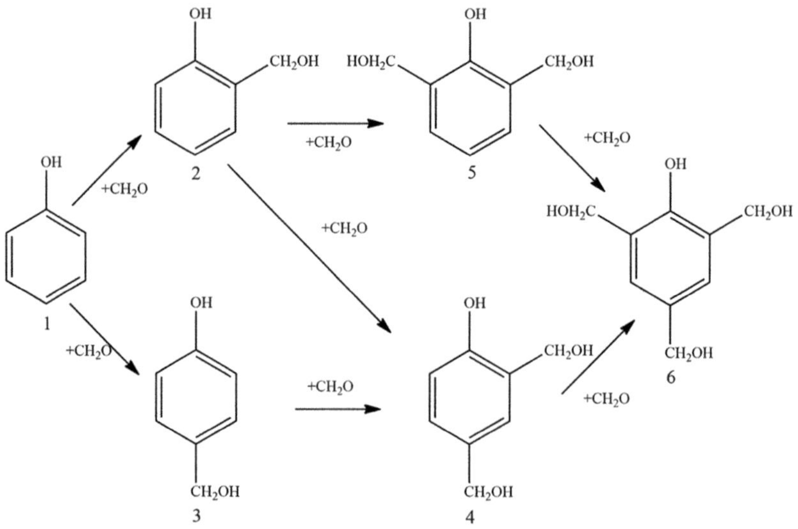

FIGURE 4.5 Structure of dimethylol phenol and trimethylol phenol.[10]

Resol-type PF consists of methylol phenol, oligomers units, and residual amounts of free phenol and formaldehyde. Reaction of resol-PF resin depends on several factors, such as molar ratio of F/P, catalyst concentration or type, and temperature. For example, at low temperature, below 60°C, only the addition of formaldehyde to phenol occurs. Meanwhile, as the temperature increases to higher than 60°C, condensation reactions of methylol phenols with phenol and/or methylol phenols occur (Fig. 4.6), leading to the production of resol-PF resin. The reaction steps for the addition and condensation conditions of resol-PF resin are summarized in Table 4.2.

TABLE 4.2 Addition and Condensation Steps for Resol Resin.[10]

	Addition	Condensation
Temperature	<60°C	>60°C
pH	8–9	9–11
Time, hours	2–4	2–4
		4–8 (pH 8–9)
Water dilutability	∞	Low
Solid content	50	80
Viscosity (mPa.s)	50–100	~5000

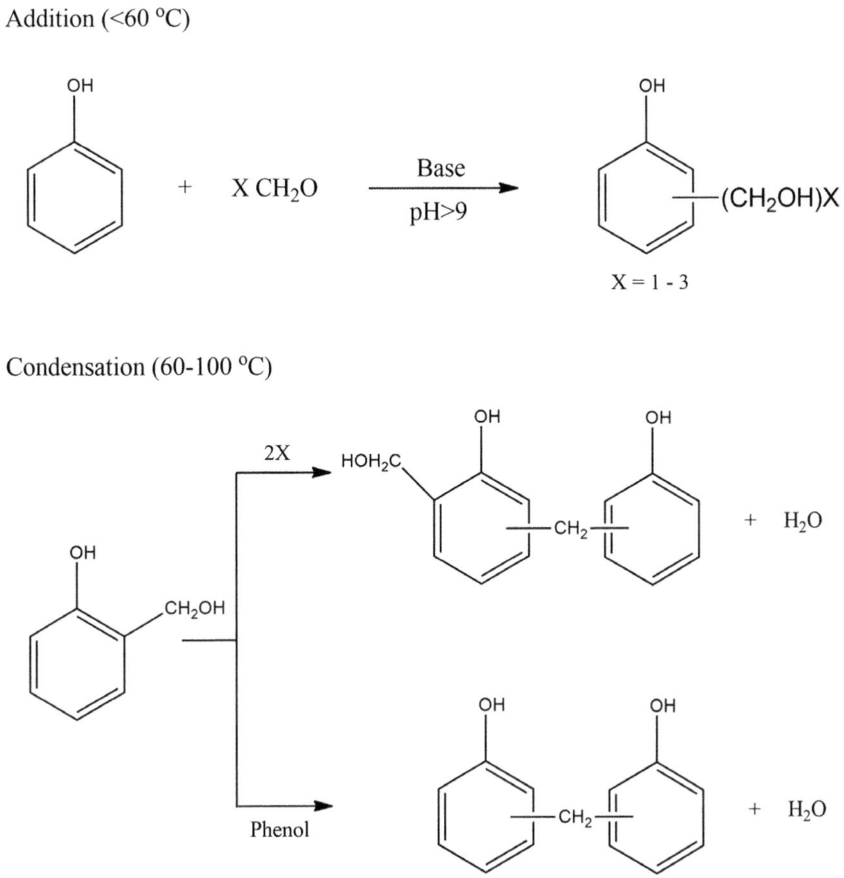

Addition (<60 °C)

HOH₂C and phenol reaction scheme

$$\text{Phenol} + X\,CH_2O \xrightarrow[\text{pH>9}]{\text{Base}} \text{(CH}_2\text{OH)X}$$

X = 1 - 3

Condensation (60-100 °C)

2X ⟶ HOH₂C ... CH₂ ... + H₂O

Phenol ⟶ CH₂ ... + H₂O

FIGURE 4.6 Addition and condensation reaction of formaldehyde to phenol.[10]

4.4 BIOMASS FEEDSTOCK FOR THE PRODUCTION OF PHENOLIC RESIN

There are increased interest and research focuses on bio-based PF resin due to the rising petroleum prices and concerns over greenhouse gas emission.[18] Many studies have been carried out in utilizing biomass as an alternative raw material to reduce the consumption of petroleum-based phenol.[19,20] Biomass is considered as one of the most popular candidates because it is the only renewable resources possessing fixed carbon, which is an important aspect in the production of conventional hydrocarbon liquid fuels and many other applications. Biomass is well recognized as the world's most abundant and promising sources to substitute petroleum-based chemicals due to its similarity in chemical structure.[21,22] It can be categorized into three main components which are cellulose, hemicellulose, and lignin.[23,24] Lignin is among the popular components because of its phenolic nature that allows various types of phenol and phenol derivatives as well as aromatic chemicals to be extracted. Among the many applications of interest in the use of lignin is the production of PF resin in which the lignin component is expected to be used to reduce the consumption of petroleum-based phenol.

The percentage of lignin in wood varies, depending on several factors such as plant species, climate, soil condition, etc. Usually it is present in between 19% and 35% of the dry weight of the plant.[25] Lignin component can be de-polymerized into monomeric and oligomeric phenolic compounds at high temperature. Hardwood and softwood are the most widely employed feedstock for the production of pyrolytic lignin due to consistency and easy availability. Many other biomass feedstocks have also been used in the production of phenolic precursors for the manufacture of phenol-based resins. An overview of the details is provided in Table 4.3 which is classified into five categories; woods, forest, industrial residues, agricultural residues, and industrial lignin.

TABLE 4.3 Biomass Feedstocks Used in the Production of Phenolic Precursors for Phenol-Based Resins.

Feedstock	Type of Feedstock
Woods	Softwood bark, ground softwood, pine, sawdust woodchips, mixed hardwoods
Forest residues	Bark waste, peat moss, treetops, limbs
Agricultural residues	Bagasse, cashew nut shell, corn bran, empty fruit bunch
Industrial residues	Lignin from newsprint, paper waste, creosote treated wood waste, birch wood waste, wood industry residues, black pulping liquor
Industrial lignins	Organosolv lignin, lignin from steam explosion of birch, lignosulphonate

Over nearly two decades of research on many types of lignin structure, it is found that lignin is a complicated amorphous polymer with three-dimensional network. Lignin is composed of three basic structural monomers, which are p-phenyl (H-type), guaiacyl monomer (G-type), and syringyl monomer (S-type), which are derived from coumaryl alcohol, coniferyl alcohol, and sinapyl alcohol, respectively. The proportion of these basic structures varies greatly in different families of plant. For an example, lignin of hardwood contains large amounts of syringyl units, while softwood lignin is mainly guaiacyl-type. Softwoods can produce more reactive phenolic compounds than hardwoods because it consists of more guaiacol unit which has one extra site available for polymerization in ortho position. Meanwhile, hardwoods consist of more syringol units and polymerization is limited only to the para position. Figure 4.7 shows the chemical structure of syringol, guaiacol, and their precursor alcohols.[24] Lignin content in softwoods is normally higher than that of hardwoods. The typical proximate composition of wood and bark is shown in Table 4.4.[24,26] The major chemical functional groups in lignin are hydroxyl, methoxyl, carbonyl, and carboxyl, in various proportions, which depend on the genetic origin and extraction processes.

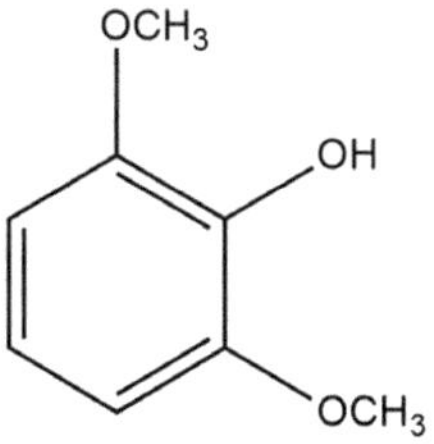

Syringol

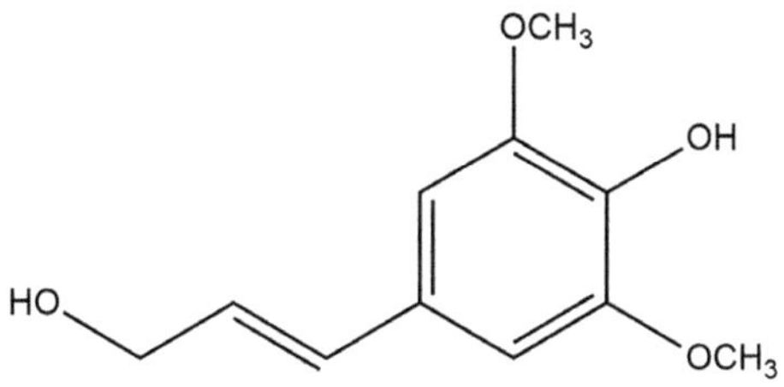

Sinapyl alcohol

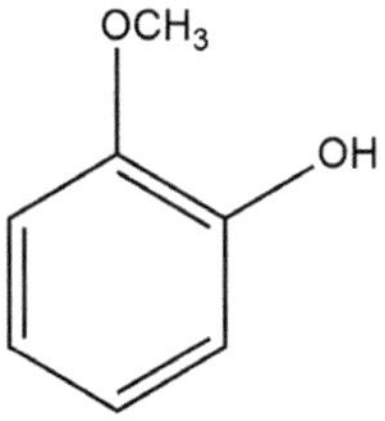

Guaiacol

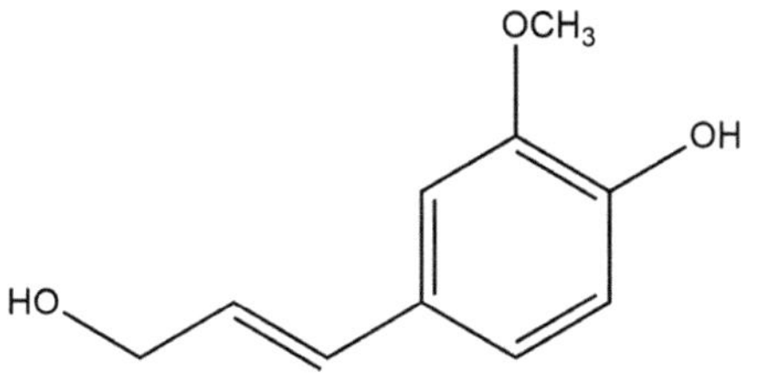

Coniferyl alcohol

FIGURE 4.7 Chemical structures of syringol and guaiacol, and their precursor alcohols (sinapyl and coniferyl alcohol).

TABLE 4.4 Typical Proximate Composition of Wood for Softwood and Hardwood.

Compound (wt%)	Softwood		Hardwood	
	Wood	Bark	Wood	Bark
Lignin	25–30	40–55	18–25	40–50
Polysaccharides (hemicelluloses and cellulose)	66–72	30–48	74–80	32–45
Extractives	2–9	2–25	2–5	5–10
Ash	0.2–0.6	Up to 20	0.2–0.6	Up to 20

4.5 PRODUCTION OF PHENOLIC COMPOUNDS FROM BIOMASS

4.5.1 FAST PYROLYSIS

Fast pyrolysis of biomass has received wide attention as it is perceived to offer significant logistical and economic advantages. The liquid product produced can be stored until required or readily transported to where it can be most effectively utilized. Fast pyrolysis is a thermal decomposition process that carried out at moderate temperature, usually 400–600°C, with a high heat transfer rate to the biomass and a short hot vapor residence time.[27] In this process, biomass is rapidly heated in the absence of oxygen and decomposed to generate mostly vapors and aerosols and some charcoal. After cooling and condensation, it transformed into a dark brown liquid which has a heating value about half of conventional fuel oil.[28] Bio-oil compounds derived from fast pyrolysis have been assorted into five main categories: phenolics, hydroxyaldehydes, hydroxyketones, sugars, and carboxylic acids.[24] Since the price and availability of synthetic PF resin are tied to that of petroleum, several attempts have been made to produce phenol or phenolic analogs from natural resources which can substitute petroleum-based phenol in the synthesis of PF resins.

Investigation on pyrolysis oils derived from bark residues, softwood, and hardwood was conducted previously.[29–32] Both novolac and resol resins were synthesized by using phenolic derivatives produced from fast pyrolysis process. The fast pyrolysis process was carried out in a small vortex reactor which was operated at 10–20 kg/h and temperature ranging from 480 to 520°C to produce primary pyrolysis oils with optimum yield (greater than 55% by weight on a dry basis). This vortex reactor transmits a very high heat flux to the biomass, causing primary de-polymerization of the biomass into polymers constituents. The pyrolysis oil was obtained after condensation of the vapors via a series of condenser at different temperatures. The first condenser was at 20°C, the second condenser at 2°C and subsequent condenser which is the third and final is at −17°C. The remaining gas and aerosol stream was passed through a coalescing filter to remove the aerosols. Meanwhile, reactive components for resin synthesis (combined phenolics and neutral fraction with high phenolic hydroxyl and aldehyde contents) were obtained by fractionating the pyrolysis oil and directly used to produce PF adhesives. The resins, i.e., novolac and resol, were evaluated and tested for wood panel production. Utilization of biomass in this invention suggests that the PF adhesives produced from the pyrolysis oil required less formaldehyde due to the presence of alde-hyde fractions in the oils. This has minimized the potential environment and health issues of excessive free formaldehyde.

Fast pyrolysis of mixed hardwoods, such as maple, birch, and beech, was performed in an air-blown bubbling fluidized-bed to produce pyrol-ysis oils asserted to be fit into production of PF resins.[33] Although Effendi et al.[24] claimed that the process is not a pure pyrolysis process because small amount of air was employed, but the oxygen only represents an order of 5% of stoichiometric combustion requirements; hence the process is considered quite close to the standard of fast pyrolysis. The pyrolysis was carried out at temperature range of 525–700°C with a gasifier residence time from 0.5 to 2.5 s to produce alkyl phenolic-rich oil. The tempera-ture in this process is well above the average temperature used by fast pyrolysis processes accepting lower oil yields but was claimed to be 100% selectivity by Himmelblau[33] and no further explanation regarding to what he meant by it. Char and ash contents in the oil obtained were lower than 1 wt% and analysis on the pyrolysis oil recognized about 40 compounds present at greater than 1% mol representing 82% mol of the material. Identified compounds are claimed to be all polymerizable with average of two free positions for methylene linkages versus three for phenol. As a result to the resins, phenol substitutions up to 50% still provide adequate bonding for water-resistant adhesives. These resins were used to produce 3-ply plywood from southern pine and their performance is comparable to commercial PF adhesives. They also reported that no cellulose compounds existed in the oil produced and the mixture is said to give phenolics with an average of two reactive sites with a large fraction of syringol would be expected.

Nakos et al. used pyrolysis oil to substitute petroleum-derived phenol in the production of resol resins.[34] The pyrolysis oil samples were obtained from different biomass feedstocks, i.e., pure wood and bark. PF resins were developed using the whole pyrolysis oil to substitute approximately 50% of phenol consumption. The pyrolysis oil can be used in the manufacture of PF resins for various panel products with positive results. The resol-type resin was successfully used in the production of oriented strand board (OSB) and plywood. The panel bond performance, however, is highly dependent not only on the resin type but also on the type of wood veneers. The results show that the bond quality of the resins produced from the pyrolysis oils is equal to or even better than that of the conventional resin.

4.5.2 VACUUM PYROLYSIS

Other than fast pyrolysis, vacuum pyrolysis has also been carried out to produce PF resin precursors from lignocellulosic materials. The difference between vacuum pyrolysis and fast pyrolysis is that the former is employed for longer residence (around 40 s). The vacuum will suppress condensation reactions in the vapor as the concentrations of reactants hence lower the reaction rates. However, this effect is not capable of fully compensating for the longer residence times, as evidenced by the fact that usually lower liquid yields are reported for vacuum pyrolysis compared to fast pyrolysis process.[24]

Roy et al. conducted vacuum pyrolysis to produce phenolic derivatives from pyrolysis oil to be used in the preparation of resol type resins.[35] Pyrolysis oils were obtained from softwood bark waste via pyrolysis process at 400–550°C under pressure of <50 kPa. The vapor collected from this process was then condensed to obtain liquid consisting phenolic-rich pyrolysis oil which has a dew point of about 65–75°C. The condensation reactions under these conditions will make the pyrolysis oils possessing fewer polymerized components and low acid content and also capable of lowering the smoky odor levels. In addition, vapor that is not condensed in the first step was condensed in a second step and additional processes such as evaporation was performed to further recover the phenolics fraction with a boiling point over 125°C. Resol-type phenolics with different molar ratio of F/P (1.75, 2.0, and 2.25) were prepared using the pyrolysis oil with 40% phenol substitution and were used for the manufacture of OSB panels. Some of the resin properties were also evaluated, including percentage of non-volatile matters, pH, gel time, and viscosity. The results showed that the pH of the produced resins was close to the pH of the commercial resin (~10.2) except for one resin synthesized with 25% phenol replacement due to having higher NaOH/P molar ratio. The resin's viscosity was also higher compared to commercial resin (90–150 cP), but within the acceptable limits of what is being used in the industry (<250 cP). Mechanical properties, such as internal bonding (IB) and torsion shear, were also higher compared to those of the control panels prepared from commercial resin.

Evaluation on vacuum pyrolysis oils as phenol substitutes in the production of resol-type PF resin was also carried out by Amen-Chen et al.[36] The pyrolysis oils were reacted directly with formaldehyde during resin synthesis without separation or fractionation processes except for the removal of low molecular weight organic acids. Production of the pyrolysis oil was performed in a pilot plant scale vacuum reactor at 505°C and 18 kPa with a feed rate of 50 kg/h. Softwood barks from balsam fir, white spruce, and black spruce were used as feedstocks. Resol-type PF resins were prepared by using 25 and 50% bark-derived oils as a replacement to the petroleum-based phenol and reacted at formaldehyde to phenol ratios of 1.75, 2.0, and 2.25. As a result, modulus of elasticity, modulus of rapture, and internal bonding of the strand boards produced by using the resins are better than those specified in the Canadian Standards; however, the thickness swelling test of the boards produced shows an opposite result. The resins produced with 25 wt% of phenol substitute were found to be suitable to be used as surface resins and the performance was as good as the commercial resin. Substitution up to 50% of phenol using the pyrolysis oil tends to reduce the internal bonding properties which can be due to the insufficient cross-linking. The resins produced from the pyrolysis oil also showed slow curing kinetics and lower thermal stability as compared to the commercial control. Attempted to improve the curing behavior of the wood adhesives was carried out by adding a small concentration of polypropylene carbonate (0.5–1.5 wt%) but no significant improvement was observed in the mechanical properties of the strand boards.[37]

4.5.3 LIQUEFACTION

Liquefaction of lignocellulosic biomass, using phenol as liquefying agent, has been intensively studied for several decades.[38] The products obtained from the liquefaction process are increasingly important to be used as starting materials for the production of new energy, solvents, fine chemicals, etc.[39,40] Most of the biomass liquefaction reactions were carried out using acid catalyst to significantly increase the conversion yield. Liquefaction under alkaline condition indeed gives low biomass residue content; however, they were not effective catalysts to achieve high amount of combined phenol.[41]

Both novolac- and resol-type PF resins have been synthesized from liquefaction of biomass. There are two different methods proposed to produce liquefied biomass-based novolac resins. The first one was to directly use liquefied wood as a novolac-like resin and the other one was to further condense liquefied biomass with formaldehyde after liquefaction. Most studies on the liquefaction of biomass were conducted to investigate the effect of liquefaction solvent to biomass ratio. The results suggested that the liquefying agent plays a double functional role during the liquefaction reaction. It reacts with the reactive sites of the biomass and also dissolves reaction intermediates and final products, thus shifting the reaction to the liquefying direction and preventing re-condensations of the decomposed biomass components. Therefore, an excessive amount of liquefying reagent is necessary to allow penetration of the liquefying reagent into biomass and to attain a satisfactory liquefaction.[42,43]

Novolac-type PF resin produced from liquefied wood has been prepared using phosphoric acid (H_3PO_4) as catalyst.[44,45] The results showed that the amount of reacted phenol during liquefaction increased as a function of liquefaction temperature, time, catalyst concentration, and phenol to wood ratio. The flowing temperature and viscosity of the resin increased with the increasing amount of phenol to wood ratio due to the cohesiveness of the liquefied wood components. A molding compound was prepared by mixing 37.7 wt% of the liquefied wood resin with 49.5 wt% wood flour as filler, 9.4 wt% hexamine as cross-linking agent, 2.4 wt% $Ca(OH)_2$ as the catalyst, and 1.0 wt% zinc sterate as release agent. The mechanical and physical properties of the molding products were evaluated and compared with commercial petroleum-based novolac resin. The flexural strength of the molding product produced from liquefied wood resin was increased with the increasing amount of reacted phenol and comparable to those of commercial control when the amount of reacted phenol reached 75%. However, water sorption test showed that the liquefied wood moldings had higher water diffusion coefficients than that of commercial novolac resin. This phenomenon suggested that the liquefied wood PF resins are more hydrophilic than the petroleum-based PF resin. The same observation was obtained in other studies using the same method but with a different catalyst, i.e., hydrochloric acid.[44,46]

Zakaria et al. produced novolac-type PF resin using liquefied oil palm empty fruit bunch (EFB) fibers.[47] They reported that flexural properties of the composite board produced using the PF resin increased as the liquefaction time increased. In terms of melt viscosity, generally the apparent melt viscosity of PEFB is higher than that of the commercial novolac resin, but with this method the apparent melt viscosity of PEFB was close to that of commercial novolac resin as the phenol/EFB ratio increased to 4. The effect of water on liquefaction and the properties of liquefied wood resin produced has also been studied by Lee and Wang.[48] The addition of water during liquefaction could somehow lower the melt viscosity of the resulted resin, and at the same time resulted in the decrease of the flexural strength and toughness of the liquefied wood resin moldings.

Resol-type PF resin is a thermoset which is synthesized by the reaction of phenol to formaldehyde at a molar ratio of >1 under alkaline conditions. Hassan et al. have studied the synthesis of resol-PF resins from phenol liquefied wood and applied the resin for particleboard.[9] Instead of using un-reacted phenol to determine the amount of formaldehyde, these authors also carried out a formaldehyde reactivity test to determine an optimum amount of formaldehyde required for the resin synthesis. In the formaldehyde reactivity test, liquefied wood was mixed with NaOH, water, and excessive amount of formaldehyde, and kept the reaction at 60°C for 5 h. The reaction samples were taken out every 30 min intervals and titrated for the determination of free formaldehyde. The amount of free formaldehyde was used to calculate the equivalent of formaldehyde reacted per 100 g of liquefied wood. Neutralized liquefied wood was then reacted with equivalent formaldehyde until the desired viscosity was achieved. The synthesized resins showed good viscosity stability, expected resin solid content, and good sprayability with a compressed air sprayer. Modulus of rupture (MOR) and modulus of elasticity (MOE) of the particleboard made from synthesized resol-PF resin are comparable with UF resin. However, the internal bonding strength of the particleboard produced using the resol-PF resin was lower than that of the UF control, which can be attributed to the incomplete curing of the PF resin in the core layer of the particleboard. As a consequence, longer hot-press time might be needed to fully cure the core layers.

The type of catalyst used during the polymerization of resol-type PF resin has been proven to have immense influence on the molecular structure of the resin. Investigation on the effect of NaOH catalyst on the properties of resol PF resins made from liquefied wood bark was carried out by Yuan et al.[7] Larch bark was liquefied with phenol at 130°C for 90 min at phenol to bark ratio 5/2 (w/w). A mixture of H_2SO_4 and H_3PO_4 was used as the catalyst at 5% (based on the phenol weight). They reported that more than 90% of the barks were successfully liquefied by using the above condition. In the same study,[7] they found that the final properties, viscosity, and reactivity of the PF resin produced can be adjusted using NaOH.

Alma and Basturk synthesized resol type PF resins from the reaction of grapevine cane *(Vitis vinisera* L.*)* liquefied using phenol under acidic condition.[20] The biomass was first liquefied with phenol in the presence of sulfuric acid as a catalyst at 150°C for 2 h. The liquefied grapevine cane powders were then resinified with different ratio of formaldehyde by using NaOH as a catalyst. Three-layer plywood were prepared by using the grapevine–cane-based PF resin and tested by using tensile-shear test following the same standard for PF resin adhesives to evaluate dry- and water-proof adhesive bond strength. The results showed the grapevine canes could easily be liquefied with phenol in the presence of sulfuric acid as a catalyst. The dry shear adhesive strength result surpasses the minimum values (12 kgf/cm²) specified in JIS K-6852 standards for the tension shear strength of resol-type adhesives. Meanwhile, the value of the wet-shear adhesives strength is found to comply with the minimum values of the JIS K-6852 which is 10 kgf/cm².

4.6 FUTURE PERSPECTIVES

Apart from the fluctuation of petroleum prices, there are also other problems that arise including the management of waste from wood and agricultural industries. Utilization of these wastes to produce high-value-added products is capable of reducing the impact of pollution to the environment and also enhancing the well-being of future development. Up to now, many studies have been reported on the utilization of biomass in the production of PF resins. However, there are still many efforts afoot to make the bio-based PF resin competitive with the commercial PF, in terms of stability, reactivity, physico-mechanical properties, production cost, sustainability, and so on.

KEYWORDS

- **phenol formaldehyde**
- **lignocellulosic biomass**
- **novolac**
- **resol**
- **phenolic precursor**
- **thermosetting resins**

REFERENCES

1. Raquez, J. M.; Deléglise, M.; Lacrampe, M. F.; Krawczak, P. Thermosetting (Bio) Materials Derived from Renewable Resources: A Critical Review. *Prog. Polym. Sci.* **2010,** *35,* 487–509.
2. Langenberg, K. V.; Grigsby, W.; Ryan, G. *Green Adhesives: Options for the Australian Industry–Summary of Recent Research into Green Adhesives from Renewable Materials and Identification of those that are Closest to Commercial Uptake;* Forest & Wood Products: Australia, 2010.
3. Schmidt, R. G.; Frazier, C. E. Network Characterization of Phenol–Formaldehyde Thermosetting Wood Adhesive. *Int. J. Adhes. Adhes.* **1998,** *18* (2), 139–146.
4. Pizzi, A. Phenolic Resin Adhesives. In *Handbook of Adhesive Technology;* 2nd ed.; Pizzi, A., Mittal, K. L., Eds.; Marcel Dekker: New York, 2003; pp 541–571.
5. Pan, H. Synthesis of Polymers from Organic Solvent Liquefied Biomass: A Review. *Renew. Sust. Energ. Rev.* **2011,** *15* (7), 3454–3463.
6. Kunaver, M.; Medved, S.; Čuk, N.; Jasiukaitytė, E.; Poljanšek, I.; Strnad, T. Application of Liquefied Wood as a New Particle Board Adhesive System. *Bioresour. Technol.* **2010,** *101* (4), 1361–1368.
7. Yuan, J.; Gao, Z.; Wang, X. M. Phenolated Larch-Bark Formaldehyde Adhesive with Various Amounts of Sodium Hydroxide. *Pigm. Resin. Technol.* **2009,** *38* (5), 290–297.

8. Vázquez, G.; Santos, J.; Freire, M.; Antorrena, G.; González-Álvarez, J. J. DSC and DMA Study of Chestnut Shell Tannins for their Application as Wood Adhesives without Formaldehyde Emission. *J. Therm. Anal. Calorim.* **2012,** *108* (2), 605–611.

9. Hassan, E. B.; Kim, M.; Wan, H. Phenol–Formaldehyde-Type Resins Made from Phenol-Liquefied Wood for the Bonding of Particleboard. *J. Appl. Polym. Sci.* **2009,** *112* (3), 1436–1443.

10. Pilato, L. Resin Chemistry. *In Phenolic Resins: A Century of Progress;* Springer Berlin: Heidelberg, 2010; pp 41–91.

11. Jones, T. T. Some Preliminary Investigations of the Phenolformaldehyde Reaction. *J. Chem. Technol. Biotechnol.* **1946,** *65* (9), 264–275.

12. Kamo, Ni.; Tanaka, J.; Higuchi, M.; Kondo, T.; Morita, M. Condensation reactions of Phenolic Resins VII: Catalytic Effect of Sodium Bicarbonate for the Condensation of Hydroxymethylphenols. *J. Wood Sci.* **2006,** *52* (4), 325–330.

13. Kamo, N.; Okamura, H.; Higuchi, M.; Morita, M. Condensation Reactions of Phenolic Resins V: Cure-Acceleration Effects of Propylene Carbonate. *J. Wood Sci.* **2004,** *50* (3), 236–241.

14. Kamo, N.; Higuchi, M.; Yoshimatsu, T.; Morita, M. Condensation reactions of Phenolic Resins IV: Self-Condensation of 2,4-Dihydroxymethylphenol and 2,4,6-Trihydroxymethylphenol (2). *J. Wood Sci.* **2004,** *50* (1), 68–76.

15. Kamo, N.; Higuchi, M.; Yoshimatsu, T.; Ohara, Y.; Morita, M. Condensation Reactions of Phenolic Resins III: Self-Condensations of 2,4-Dihydroxymethylphenol and 2,4,6-Trihydroxymethylphenol (1). *J. Wood Sci.* **2004,** *48* (6), 491–496.

16. Higuchi, M.; Yoshimatsu, T.; Urakawa, T.; Morita, M. Kinetics and Mechanisms of the Condensation Reactions of Phenolic Resins II. Base-Catalyzed Self-Condensation of 4-Hydroxymethylphenol. *Polymer.* **2001,** *33* (10), 799–806.

17. Higuchi, M.; Urakawa, T.; Morita, M. Condensation Reactions of Phenolic Resins. 1. Kinetics and Mechanisms of the Base-Catalyzed Self-Condensation of 2-Hydroxymethylphenol. *Polymer.* **2001,** *42* (10), 4563–4567.

18. Sulaiman, O.; Salim, N.; Hashim, R.; Yusof, L. H. M.; Razak, W.; Yunus, N. Y. M.; Hashim, W. S.; Azmy, M. H. Evaluation on the Suitability of Some Adhesives for Laminated Veneer Lumber from Oil Palm Trunks. *Mater. Design.* **2009,** *30* (9), 3572–3580.

19. Roslan, R.; Zakaria, S.; Chia, C. H.; Boehm, R.; Laborie, M. P. Physico-Mechanical Properties of Resol Phenolic Adhesives Derived from Liquefaction of Oil Palm Empty Fruit Bunch Fibres. *Ind. Crop. Prod.* **2014,** *62,* 119–124.

20. Alma, M. H.; Basturk, M. A. Liquefaction of Grapevine Cane (*Vitis vinisera* L.) Waste and its Application to Phenol–Formaldehyde Type Adhesive. *Ind. Crop. Prod.* **2006,** *24* (2), 171–176.

21. Zakaria, S.; Roslan, R.; Amran, U. A.; Chia, C. H.; Bakaruddin, S. B. Characterization of Residue from EFB and Kenaf Core Fibers in the Liquefaction Process. *Sains Malays.* **2014,** *43* (3), 429–435.

22. El Mansouri, N. E.; Yuan, Q.; Huang, F. Study of Chemical Modification of Alkaline Lignin by the Glyoxalation Reaction. *Bioresources.* **2011,** *6* (4), 4523–4536.

23. Amran, U. A.; Zakaria, S.; Chia, C. H.; Jaafar, S. N. S.; Roslan, R. Mechanical Properties and Water Absorption of Glass Fibre Reinforced Bio-Phenolic Elastomer (BPE) Composite. *Ind. Crop. Prod.* **2015,** *72,* 54–59.

24. Effendi, A.; Gerhauser, H.; Bridgwater, A. V. Production of Renewable Phenolic Resins by Thermochemical Conversion of Biomass: A Review. *Renew. Sust. Energ. Rev.* **2008,** *12* (8), 2092–2116.
25. Mohan, D.; Pittman, C. U.; Steele, P. H. Pyrolysis of Wood/Biomass for Bio-Oil: A Critical Review. *Energ. Fuel.* **2006,** *20* (3), 848–889.
26. Scholze, B.; Meier, D. Characterization of the Water-Insoluble Fraction from Pyrolysis Oil (Pyrolytic Lignin). Part I. PY–GC/MS, FTIR, and Functional Groups. *J. Anal. Appl. Pyrol.* **2001,** *60* (1), 41–54.
27. Czernik, S.; Bridgwater, A. V. Overview of Applications of Biomass Fast Pyrolysis Oil. *Energ. Fuel.* **2004,** *18* (2), 590–598.
28. Bridgwater, A. V.; Meier, D.; Radlein, D. An Overview of Fast Pyrolysis f Biomass. *Org. Geochem.* **1999,** *30* (12), 1479–1493.
29. Chum, H.; Black, S.; Diebold, J.; Kreibich R. E. Resole Resin Products Derived from Fractionated Organic and Aqueous Condensates Made by Fast-Pyrolysis of Biomass Materials. U.S. Patent 5, 235, 021, 1993.
30. Chum, H.; Black, S.; Diebold, J.; Kreibich R. E. Phenolic Compounds Containing Neutral Fractions Extract and Products Derived therefrom from Fractionated Fast-Pyrolysis Oils. U.S. Patent 5, 223, 601, Midwest Research Institute Ventures: US, 1993.
31. Chum, H.; Kreibich, R. E. Process for Preparing Phenolic Formaldehyde Resole Resin Products Derived from Fractionated Fast-Pyrolysis Oils. U.S. Patent 5. 091. 499, Midwest Research Institute: US,1992.
32. Chum, H.; Diebold, J.; Scahill, J.; Johnson, D.; Black, S.; Schroeder, H.; Kreibich R. E. In *Biomass Pyrolysis Oil Feedstocks for Phenolic Adhesives.* ACS Symposium Series, American Chemical Society: Washington, D.C, 1989; Vol. 385, pp 135–151.
33. Himmelblau, A. Method and Apparatus for Producing Water-Soluble Resin and Resin Product Made by that Method. U.S. Patent 5, 034, 498, Biocarbons Corporation: 1991.
34. Nakos, P.; Tsiantzi, S.; Athanassiadou, E. In *Wood Adhesives Made with Pyrolysis Oils.* Proceedings of the 3rd European Wood-Based Panel Symposium, Hanover, Germany, Sept 12–14, 2001; European Panel Federation and Wilhelm Klauditz Institute: Germany, 2001.
35. Roy, C.; Liu. X.; Pakdel, H. Process for the Production of Phenolic-Rich Pyrolysis Oils for Use in Making Phenol–Formaldehyde Resole Resins. U.S. Patent 6, 143, 856, Pyrovac Technologies Inc.: 2000.
36. Amen-Chen, C.; Riedl, B.; Wang, X. M.; Roy, C. Softwood Bark Pyrolysis Oil-PF Resols. Part 1. Resin Synthesis and OSB Mechanical Properties. *Holzforschung.* **2002,** *56,* 167.
37. Amen-Chen, C.; Riedl, B.; Wang, X. M.; Roy, C. Softwood Bark Pyrolysis Oil-PF Resols. Part 3. Use of Propylene Carbonate as Resin Cure Accelerator. *Holzforschung.* **2002,** *56,* 281.
38. Niu, M.; Zhao, G. J.; Alma, M. Polycondensation Reaction and its Mechanism During Lignocellulosic Liquefaction by an Acid Catalyst: A Review. *Forestry Stud. China.* **2011,** *13* (1), 71–79.
39. Demirbas, M. F. Conversion of Corn Stover to Chemicals and Fuels, Energy Sources, Part A: Recovery, Utilization, and Environmental Effects. *Energ. Source.* **2008,** *30,* 788–796.

40. Demirbas, M. F.; Balat, M. Biomass Pyrolysis for Liquid Fuels and Chemicals: A Review. *J. Sci. Ind. Res. India.* **2007,** *66,* 797–804.
41. Alma, M. H.; Maldas, D.; Shiraishi, N. Liquefaction of Several Biomass Wastes into Phenol in the Presence of Various Alkalis and Metallic Salts as Catalysts. *J. Polym. Eng.* **1998,** *18,* 161.
42. Bakarudin, S. B.; Zakaria, S.; Chia, C. H.; Jani, S. M. Liquefied Residue of Kenaf Core Wood Produced at Different Phenol-Kenaf Ratio. *Sains Malays.* **2012,** *41* (2), 225–231.
43. Pan, H.; Shupe, T. F.; Hse, C. Y. Characterization of Liquefied Wood Residues from Different Liquefaction Conditions. *J. Appl. Polym. Sci.* **2007,** *105* (6), 3740–3746.
44. Lin, L.; Yoshioka, M.; Yao, Y.; Shiraishi, N. Physical Properties of Moldings from Liquefied Wood Resins. *J. Appl. Polym. Sci.* **1995,** *55* (11), 1563–1571.
45. Lin, L.; Yoshioka, M.; Yao, Y.; Shiraishi, N. Liquefaction of Wood in the Presence of Phenol Using Phosphoric Acid as a Catalyst and the Flow Properties of the Liquefied Wood. *J. Appl. Polym. Sci.* **1994,** *52* (11), 1629–1636.
46. Alma, M.; Yoshioka, M.; Yao, Y.; Shiraishi, N. Preparation and Characterization of the Phenolated Wood Using Hydrochloric Acid (Hcl) as a Catalyst. *Wood. Sci. Technol.* **1995,** *30* (1), 39–47.
47. Zakaria, S.; Ahmadzadeh, A.; Roslan, R. Flow Properties of Novolak-Type Resin Made From Liquefaction of Oil Palm Empty Fruit Bunch (EFB) Fibres Using Sulfuric Acid as a Catalyst. *Bioresources.* **2013,** *4* (8), 5884–5894.
48. Lee, S. H.; Wang, S. Effect of Water on Wood Liquefaction and the Properties of Phenolated Wood. *Holzforschung.* **2005,** *59,* 628.

CHAPTER 5

IMPEDANCE SPECTROSCOPY: A PRACTICAL GUIDE TO EVALUATE PARAMETERS OF A NYQUIST PLOT FOR SOLID POLYMER ELECTROLYTE APPLICATIONS

SITI ROZANA BT. ABDUL KARIM[1], CHIN HAN CHAN[1,*], and LAI HAR SIM[2]

[1]*Faculty of Applied Sciences, Universiti Teknologi MARA, Shah Alam 40450, Malaysia*

[2]*Centre of Foundation Studies, Universiti Teknologi MARA, Puncak Alam 42300, Malaysia*

*Corresponding author. E-mail: cchan@salam.uitm.edu.my

CONTENTS

Abstract 98

5.1 Overview 98

5.2 Nyquist Plot 99

5.3 Bulk Resistance (R_b) Determintation from the Nyquist Plot 99

5.4 Estimation of Debye Relaxation Exponent, N 121

5.5 Summary 127

5.6 Acknowledgments 127

Keywords 127

References 127

ABSTRACT

Ionic conductivity (σ_{DC}) is one of the most important part in solid polymer electrolyte systems as it will give better understanding of the ion mobility inside polymer chain. Therefore, the determination of bulk resistance (R_b) from the Nyquist plot is crucial. Traditionally, the value of R_b is extracted using the graphical method; however, the possibility to have inconsistent value of σ_{DC} is high. Thus, in this chapter, the newly developed mathematical method using the Origin® software is introduced. The step-by-step guidelines to evaluate the R_b, σ_{DC} and Debye relaxation exponent, n using PEO/PMMA system as sample work are discussed in detail.

5.1 OVERVIEW

Impedance spectroscopy (IS) is the most popular approach in electrical measurement to elucidate the electrochemical behavior of an electrode or electrolyte. This technique involves the application of an alternating voltage to produce a steady flow of current through the sample without forming any polarization cloud at the electrodes, thus, eliminating the possibility of creating a concentration gradient within the electrolyte. The basic concepts and applications of IS have been discussed in detail.

In this chapter, a practical guideline is given in the form of guided procedures with appropriate illustrations to evaluate various parameters using information extracted from the Nyquist plots for solid polymer electrolytes (SPEs), where electric bulk properties of the film might be represented by a capacitor C in parallel with the resistance R. Besides, precautionary measures to minimize possible errors during the analysis are thoroughly discussed. This chapter starts with the introduction of the different types of Nyquist plots and the application of graphical method to determine the bulk resistance (R_b) from each type of Nyquist plot, with illustrations where necessary. The graphical method introduced here is simpler, more user friendly and serves as an alternative to that applied by Winie and Arof. In addition, a mathematical approach on evaluating the values of R_b, ionic conductivity (σ_{DC}), and the Debye relaxation exponent n is systematically presented and described in depth. Origin® software is used throughout the chapter especially in curve fitting of the Nyquist plot using the mathematical method.

Impedance Spectroscopy: A Practical Guide

5.2 NYQUIST PLOT

SPE is a thin film consisting of an ionic salt mixed with an appropriate polymer which allows ion conduction. The polymer and the ionic salt in the mixture are not in equilibrium, which means the mixture is a heterogeneous system consisting of polymer rich and polymer poor micro-phases. From the electrical perspective, the system exhibits both the resistive and the capacitive characteristics in which the resistive component represents the real (Z') part while the capacitive component denotes the imaginary (Z'') part of the complex impedance (Z^*). The relationship of Z' and Z'' can be expressed in eq 5.1 below:

$$Z^* = Z' - iZ''. \tag{5.1}$$

The measured Z' and Z'' of the impedance data are plotted in a frequency dispersion curve known as the complex *alternating current* (AC)–impedance spectrum of the electrolyte or frequently called the Nyquist plot with both the axes having the same scale. Figures 5.1(a)–(d) depict the various forms of Nyquist plot (Z' vs. Z''). Each point in the plot corresponds to the impedance of the electrolyte measured at a particular frequency. The semicircle in the higher frequency range of Figure 5.1(b) indicates the bulk impedance of the polymer electrolyte whereas the low frequency inclined spike (tail-like) is related to the interfacial impedance. On the whole, a Nyquist plot gives an insight into the electrical behavior of a material without reflecting on its direct frequency response to an applied field.

5.3 BULK RESISTANCE (R_b) DETERMINTATION FROM THE NYQUIST PLOT

5.3.1 EXTRACTING THE R_b USING THE GRAPHICAL METHOD

When a small alternating voltage is applied to the freestanding thin film of the polymer–salt system sandwiched between two block electrodes, a steady flow of current is produced through the SPE. The frequency analyzer is capable of measuring directly the real and the imaginary components of impedance using complex number, over a wide range of frequencies. The value of σ_{DC} can be evaluated from the R_b value of the

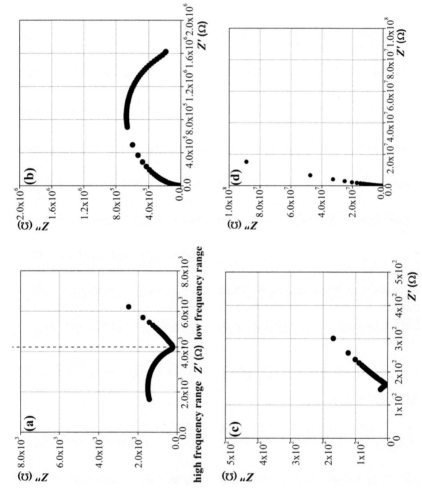

FIGURE 5.1 Various forms of Nyquist plot obtained for PEO/PMMA/LiClO$_4$ blend electrolyte.

electrolyte film obtained from the Nyquist plot using semicircle fitting based on the analysis and computer program of Boukamp.[4,5] The intersection between the semicircle and the real impedance axis (the *x*-axis) of the Nyquist plot marks the R_b of the electrolyte.[6-8] The procedures apply to extract the R_b values from the different forms of Nyquist plot are discussed in the following sections.

5.3.1.1 NYQUIST PLOTS WITH A SEMICIRCLE

Figure 5.2 displays a Nyquist plot with an incomplete circle followed by an inclined spike and the value of R_b can be obtained by completing the semicircle of the Nyquist plot. However, in order to complete the semicircle, one needs to locate the center of the circle. This can be done by drawing two chords initially on the semicircle (the positions of the two chords should cover the range of frequencies from relatively high to relatively low frequency regions of the semicircle) as shown in Figure 5.3, then a normal is drawn to each of the two chords using a divider. The center is the point of intersection between the two normal lines. This method is also applied to Nyquist plots of the type presented in Figure 5.4 which depicts an incomplete semicircle without an inclined spike. Samples with relatively low conductivity usually exhibit this type of Nyquist plot.

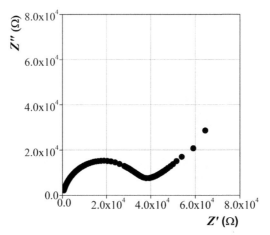

FIGURE 5.2 Nyquist plot of PEO/PMMA 75/25 doped with 10 wt% LiClO$_4$ depicting a near perfect semicircle followed by an inclined spike.

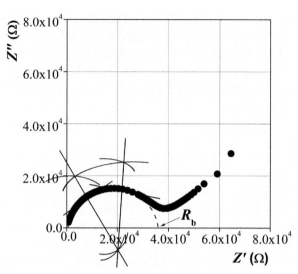

FIGURE 5.3 Schematic drawing of two normal lines on the semicircle of the Nyquist plot for PEO/PMMA 75/25 blend doped with 10 wt% $LiClO_4$.

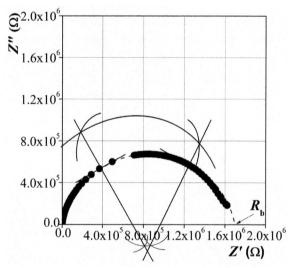

FIGURE 5.4 Locating the centetre of a semicircle without an inclined spike for PEO/PMMA 50/50 doped with 7 wt% $LiClO_4$.

For Nyquist plot of the type shown in Figure 5.5 in which a short portion of the semicircle appears before the inclined spike, only one chord could be drawn. In this case, the center of the circle can only be fixed by trial and error, such that the semicircle drawn must connect the origin and the short portion of the semicircle and intersect with the abscissa.

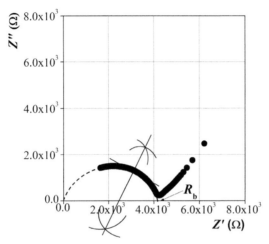

FIGURE 5.5 Nyquist plot showing a short portion of the semicircle followed by an inclined spike for PEO doped with 20 wt% LiClO$_4$.

5.3.1.2 NYQUIST PLOT WITHOUT A SEMICIRCLE

Theoretically, when the impedance of a sample of SPE is determined at high temperature especially above the glass transition temperature (T_g), the Li$^+$ ions at the electrolyte–electrode interface will move faster as the polymer chain become more flexible and in liquid-like state, thus showing high σ_{DC} value. In this case, the semicircle is completely absent with only the inclined spike or an inclined spike that looks like a "tick" appearing at the lower frequency region some distance away from the origin of the Nyquist plot as shown in Figure 5.6. For this type of Nyquist plot, a linear curve is drawn following the tangent of the top few impedance data at the low frequency region, until it intersects with the real part (Z') of the plot or the abscissa and the point of intersection is the R_b value.

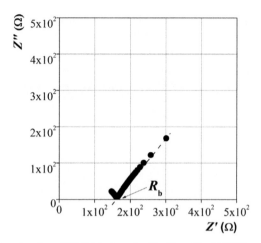

FIGURE 5.6 Nyquist plot of PMMA doped with 25 wt% of LiClO$_4$ at 120°C showing only an inclined spike without the semicircle.

Figure 5.7 depicts Nyquist plots of SPE samples that exhibit very low ionic conductivity. This type of Nyquist plot usually displays an inclined spike starting from the origin but either forming a curve toward the low frequency region as in Figure 5.7(a) or stays close to the y-axis as shown in Figure 5.7(b).

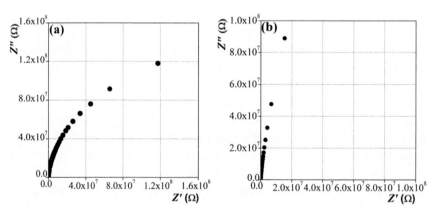

FIGURE 5.7 Nyquist plots of samples with very low conductivities. For example, PMMA doped with (a) 25 wt% and (b) 7 wt% LiClO$_4$.

Impedance Spectroscopy: A Practical Guide   105

The initial step in the determination of the R_b value is to increase the scales on both the axes as shown in Figure 5.8. Nevertheless, the extent to which the scale is increased has to be done on a trial and error basis depending on the individual Nyquist plot.

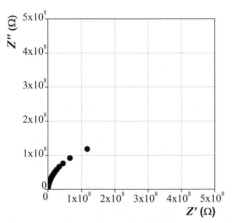

FIGURE 5.8 Expanded scale of Nyquist plot for PMMA doped with 25 wt% $LiClO_4$.

After which, two chords are drawn on the remaining impedance data followed by drawing a normal line to each chord and the point of intersection between the two normal lines as shown in Figure 5.9 marks the center of the semicircle. The detailed description of the procedure can be

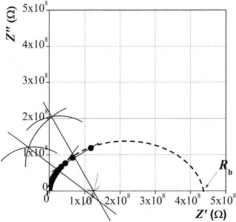

FIGURE 5.9 Schematic diagram showing intersection of the normal lines drawn on the expanded scale of the Nyquist plot for PMMA doped with 25 wt% $LiClO_4$.

found in Section 5.3.1.1. The R_b value can be obtained by completing the semicircle within the limits of the scales of the two axes to intersect with the abscissa. In fact, low conductivities ($<10^{-11}$ S cm^{-1}) obtained using this method is only an approximation as the conductivity may be influenced by the air trapped in between the electrode–electrolyte gap.

5.3.2 EVALUATION OF THE R_b VALUE USING THE MATHEMATICAL METHOD

Another approach to determine R_b other than the graphical method is to apply the mathematical method. The mathematical method allows one to analyze the various forms of Nyquist plot more systematically and accurately, thereby, enables one to get a more precise R_b value. For the mathematical method, the Origin® software (version Pro 8.1) is selected to perform all the mathematical operations including the plotting of graphs. The application of Origin® software version is possible, except certain procedures may have to be adjusted. The mathematical method of extracting the R_b value and other important electrical data of selected SPE samples is presented in the form of guided procedure in each subsection that follows.

5.3.2.1 PLOTTING THE GRAPH OF Z', Z'' VERSUS FREQUENCY (f)

The Nyquist plot of Z' vs. Z'' for polymer electrolytes can be represented by a circuit model consisting of resistors and capacitors and is frequently known as the resistor–capacitor circuit (RC circuit). This subsection is devoted to determine the R_b value precisely by plotting the Z' and Z'' impedance data of the SPE sample against the frequency (f) at which the impedance data are measured. The first step of this procedure is to extract the values of f, Z', and Z'' from the Excel file of the particular Nyquist plot of the sample as shown in Figure 5.10(a) and paste the values in the data file of the Origin® software as shown in Figure 5.10(b). After which, a "Scatter" plot type is chosen with the values of f in the x-axis and those of Z' and Z'' in the y-axis. The logarithm scale (log to the base of 10) is selected for both the two axes.

Impedance Spectroscopy: A Practical Guide 107

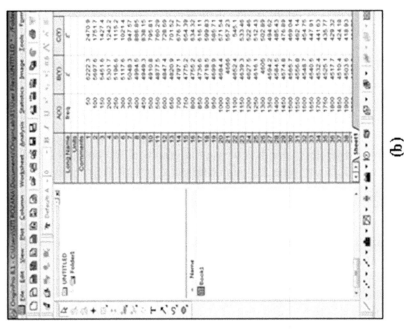

(a)

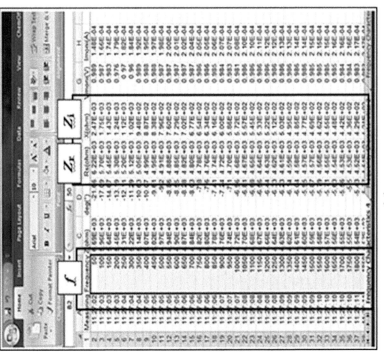

(b)

FIGURE 5.10 (a) The Excel and (b) the Origin® data files of PEO/PMMA 75/25 doped with 10 wt% LiClO$_4$ system.

Generally, two different types of plots can be obtained depending on the ionic conductivity of the polymer electrolyte sample examined.

In Figure 5.11(a), no intersection between the values of Z' and Z'' is observed over the range of frequency where the impedance data of the sample is measured. This type of plot is represented by Nyquist plots of the type shown in Figure 5.7(b) or may also be Figure 5.7(a) in Section 5.3.1.2 implying that the ionic conductivity of the sample is very low. Unfortunately, the mathematical method is not designed for samples of very low conductivity, therefore, one has to use the graphical method described in Section 5.3.1.2 to estimate the approximate R_b values for samples which exhibit Nyquist plots shown in Figures 5.7(a) and (b).

On the contrary, intersection between values of Z' and Z'' is observed in Figure 5.11(b) indicates that the electrolyte sample under investigation exhibits relatively high σ_{DC} value. The Nyquist plots for this type of samples are most probably of the type demonstrated by Figures 5.3–5.6 in Section 5.3.1.2. The impedance data of polyethylene oxide/poly methyl methacrylate (PEO/PMMA) 75/25 blend doped with 10 wt% of LiClO$_4$ gives rise to Nyquist plot of a semicircle followed by an inclined spike as shown in Figure 5.12(a) and the double logarithm plot between Z', Z'', and f as shown in Figure 5.12(b) is similar to that shown in Figure 5.11(b).

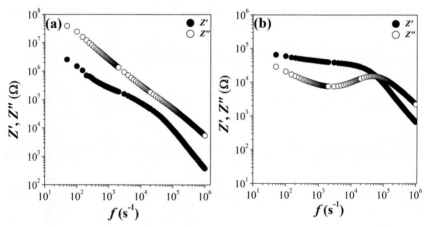

FIGURE 5.11 The plots of Z', Z'' versus f for (a) PEO/PMMA 25/75 doped with 5 wt% LiClO$_4$ and (b) PEO doped with 12 wt% LiClO$_4$.

Impedance Spectroscopy: A Practical Guide

From the plot in Figure 5.12(b), the peak of the Z'' curve is marked and identified as Z''_{max}. After which, a horizontal line and a vertical line are drawn parallel to the x- and y-axes, respectively. The point of intersection between the vertical line and the x-axis is given the symbol (f_o) which is the frequency at which the Z'' value reaches Z''_{max}. On the other hand, the y-intercept between the y-axis and the horizontal line is equal to $R_b/2$. The frequency f_o and the value of $R_b/2$ are presented in Figure 5.12(a). From the procedure described above, it is crucial for one to get the exact value of Z''_{max} in order to determine the R_b value more precisely. The procedure to locate the exact position of Z''_{max} on the peak of the Z'' curve is explained in detail in the next section.

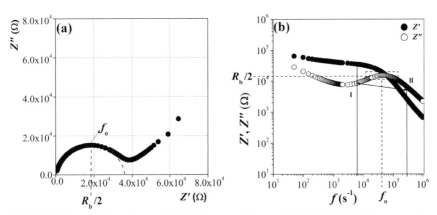

FIGURE 5.12 (a) Nyquist and (b) Z', Z'' versus f plots of PEO/PMMA 75/25 doped with 10 wt% $LiClO_4$.

5.3.2.2 DETERMINATION OF THE PRECISE VALUE OF R_b VIA POLYNOMIAL CURVE FITTING

The first step of the procedure is to re-plot the double logarithm graph of Z', Z'' vs. f by selecting data points in the vicinity of the peak, 30–40 data points from point (I) to point (II) for both the frequency dependent Z' and Z'' curves as shown in Figure 5.12(b). The resulting graph shown in Figure 5.13 depicts that the peak of the Z'' curve and its immediate vicinity has been expanded. The second step is to use the same data points as the two curves in Figure 5.13 and plot separately Z' and Z'' as functions of f with linear scale as demonstrated in Figure 5.14.

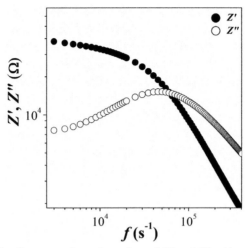

FIGURE 5.13 The frequency dependent plots of Z' and Z'' with expanded logarithm scales.

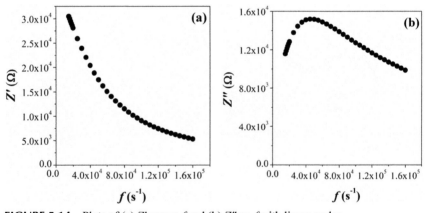

FIGURE 5.14 Plots of (a) Z' versus f and (b) Z'' vs. f with linear scales.

The third step is to perform curve fitting on the two separate curves in Figure 5.14 using the polynomial functions. The data points of Z' and the corresponding data points of Z'' are chosen to perform the curve fitting by second and fourth orders (other orders may be applicable as well) of the polynomial function, respectively. Figure 5.15 presents the commands selected in the polynomial fit dialog box in order for the Origin® software to perform the polynomial fitting of the curves.

Impedance Spectroscopy: A Practical Guide

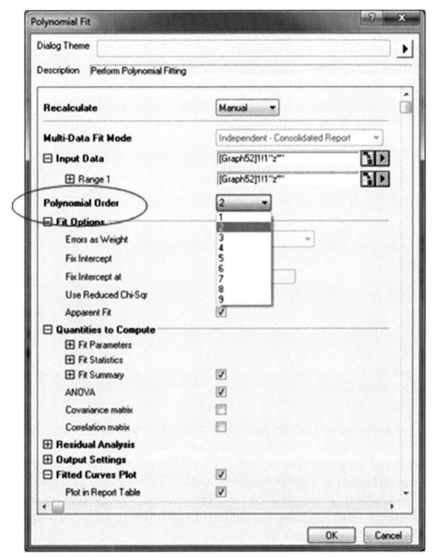

FIGURE 5.15 Command for curve fitting in Origin software.

After the curve-fitting process is completed, the results of the polynomial fitting for each curve are displayed in a table as shown in the insets in Figures 5.16 (a) and (b). The results shown in each table include the equation of the polynomial function, the correlation factor (R^2), the y-intercept, and the coefficients of a series of variable x raised to different powers depending on the order of the polynomial function. It is worthwhile to check the R^2 value to make sure a good fit between the selected data points and the chosen order of the polynomial function. A good curve fitting usually gives R^2 in the range of 0.970–0.999. Figures 5.16(a) and (b) show that both the Z' and Z'' curves perfectly fit into the regression curves of the second and fourth order polynomial functions, respectively, for the PEO/PMMA 75/25 blend doped with 10 wt% of LiClO$_4$.

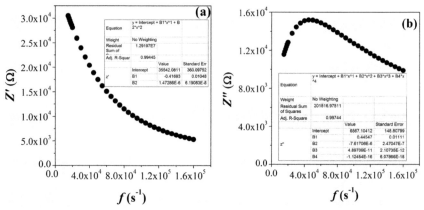

FIGURE 5.16 Curve fitting of (a) Z' and (b) Z'' using second and fourth order polynomial function, respectively.

Once good curve fittings are attained, the next move is to get the value of f_o from the results of the polynomial curve fitting. The following procedure is applied to evaluate the value of f_o systematically. The initial step is to extract the fourth order polynomial equation from the tabulated result of the plot, Z'' vs. f as shown in eq 5.2. This equation is taken from the inset of Figure 5.16(b) which belongs to the PEO/PMMA 75/25 doped with 10 wt% LiClO$_4$.

From Figure 5.16(b):

$$Z''(f) = 6567.1 + 0.45f - 7.62 \times 10^{-6} f^2 + 4.90 \times 10^{-11} f^3 - 1.12 \times 10^{-16} f^4. \quad (5.2)$$

To get the maximum value of Z'',

$$\frac{d(Z'')}{df} = 0. \quad (5.3)$$

Impedance Spectroscopy: A Practical Guide

To differentiate eq 5.2

$$\frac{d(Z'')}{df} = 0.45 - 1.52 \times 10^{-5} f + 1.47 \times 10^{-10} f^2 - 4.48 \times 10^{-16} f^3. \quad (5.4)$$

The differentiated regression curve of Z'' in eq 5.4 is equated to zero as in eq 5.3

$$0.45 - 1.52 \times 10^{-5} f + 1.47 \times 10^{-10} f^2 - 4.48 \times 10^{-16} f^3 = 0. \quad (5.5)$$

The Origin® software is applied to solve the above cubic equation to get the value of f_o. Simple instructions are given here to assist readers of this chapter to use the Origin® software to plot a function graph. The initial step is to click "Plot" then select "Scatter" without filling in any values to the data file. This operation allows one to get an empty graph with only the axes labeled with the same scale. The next step is to click "Graph" and select "Add function graph." This operation brings out a dialog box as shown in Figure 5.17, insert eq 5.5 into the column F27(x) in the Plot Detail dialog box by writing the equation in computer language as given in eq 5.6:

$$0.45 - (1.52e - 5 * x) + (1.47e - 10 * x^2) - (4.48e - 16 * x^3). \quad (5.6)$$

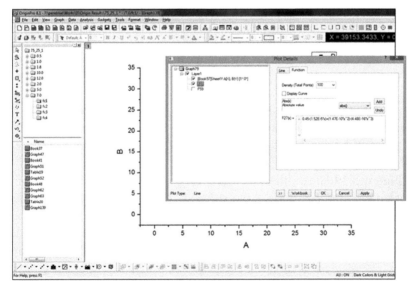

FIGURE 5.17 The appearance of the graph to which the cubic eq 5.5 is inserted and the Plot Detail dialog box on top of the empty graph.

The finishing step is to click "Apply" and "Ok" and a graph consists of a horizontal line parallel to the x-axis appears as shown in Figure 5.18.

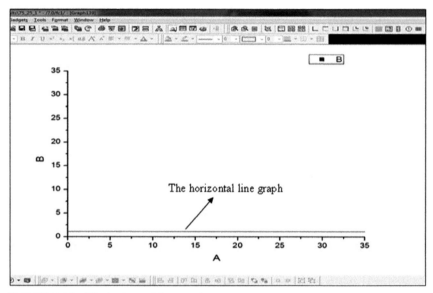

FIGURE 5.18 The horizontal line graph.

Now, rescale the x- and y-axes in such a way that the horizontal line curve slopes downwards and intersects with the x-axis (i.e., at $y = 0$) as shown in Figure 5.19. The intersection point is the value of f_o, the frequency at which Z'' reaches Z''_{max}. It is crucial to ensure that both the x- and y-axes must start from zero. One must zoom in and expanded at the point of intersection until the value of f_o at $y = 0$ can be clearly read and recorded.

Figure 5.19 shows that the value of $f_o = 50000$ s^{-1}. The value of Z' is calculated by substituting the value of f_o into the quadratic equation extracted from the result tabulated for the regression curve of Z' as shown in Figure 16(a). The value of R_b can then be determined from the equation $Z' = R_b/2$ (c.f. Section 5.3.2.1). The calculation described above is shown below:

Impedance Spectroscopy: A Practical Guide

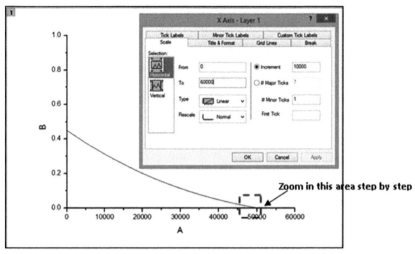

FIGURE 5.19 Adjusting the scales of the two axes to locate the exact value of f_o.

$Z'(f) = 35542.08 - 0.42f + 1.47 \times 10^{-6} f^2$ in which $R^2 = 0.994$
$Z'(f_o) = 35542.08 - 0.42(50000) + 1.47 \times 10^{-6} (50000)^2$
$\quad\quad = 1.82 \times 10^4 \: \Omega$
$Z'(f_o) = R_b / 2$
$\quad R_b = 2 \: Z_r(f_o)$
$\quad\quad = 3.64 \times 10^4 \: \Omega.$

From the regression curve of Z'', the value of Z''_{max} at frequency f_o is calculated by substituting the value of f_o into eq 5.2.

$Z''(f) = 6567.1 + 0.45f - 7.62 \times 10^{-6} f^2 + 4.90 \times 10^{-11} f^3 - 1.12 \times 10^{-16} f^4$ where $R^2 = 0.997$
$Z''(f_o) = 6567.1 + 0.45(50000) - 7.62 \times 10^{-6} (50000)^2 + 4.90 \times 10^{-11} (50000)^3 - 1.12 \times 10^{-16} (50000)^4$
$Z''(f_o) = 1.54 \times 10^4 \: \Omega.$

5.3.2.3   TO DETERMINE THE DC ION CONDUCTIVITY (σ_{DC}) OF SPE

The DC conductivity (σ_{DC}) of an SPE can be evaluated using eq 5.7

$$\sigma_{DC} = L/R_b A. \quad\quad\quad (5.7)$$

The symbol L denotes the thickness of the polymer electrolyte film which is usually an average of at least three measurements at three different locations of the film using a digital caliper while A is the surface area of the film in contact with the surface of the electrode used during the impedance measurement. The example below demonstrates the determination of σ_{DC} for the PEO/PMMA 75/25 blend doped with 10 wt% of LiClO$_4$.

Example

The sample is scanned over the range of frequency, 50 Hz–1 MHz and a total of 4–6 measurements are taken for each sample. The R_b value for each measurement is obtained either using the graphical method as described in Section 5.3.1 or by the mathematical method as explained in detail in Sections 5.3.2.1 and 5.3.2.2.

The R_b value of the sample = 3.64 × 10^4 Ω.

The average thickness (L) of the sample film measured using a Mitutoyo Digimatic Calliper = 0.031 cm.
Diameter of the electrode = 2.00 cm.
(Note: before analysis, it is advisable to check the diameter of the electrode).

Radius of the electrode = 1.00 cm.

Therefore, surface area of the sample film = surface area of the electrode
$$= \pi r^2$$
$$= 3.142 \text{ cm}^2$$

$$\sigma_{DC} = L/R_b A$$
$$= 0.031 \text{ cm}/3.64 \times 10^4 \ \Omega \times 3.142 \text{ cm}^2$$
$$= 2.68 \times 10^{-6} \text{ S cm}^{-1}$$

$$\Delta\sigma_{DC} = \left(\frac{\Delta L}{L} + \frac{\Delta R_b}{R_b} \right) \sigma_{DC}. \tag{5.8}$$

Equation 5.8 is employed to estimate the percent error of σ_{DC} as shown in Tables 5.1 and 5.2. Symbols ΔR_b, ΔL, and σ_{DC} are standard deviations of

R_b, L, and σ_{DC} values, respectively. Table 5.3 displays the comparison of σ_{DC} and the error of σ_{DC} for PEO/PMMA 25/75 with addition of LiClO$_4$. From the table, one notices that the mean σ_{DC} determined using the graphical and the mathematical methods is very close as evident in the percentage difference in σ_{DC} value between the two methods for the samples are approximately less than 10% except for sample with $Y_S = 15.0$ wt%.

TABLE 5.1 The Standard Deviations ΔL, ΔR_b, and $\Delta \sigma_{DC}$ for PEO/PMMA 25/75 Doped with 12 wt% LiClO$_4$ Calculated after eq 5.8 Where the R_b Value Is Obtained Using the Mathematical Method.

Mass Ratio of LiClO$_4$, Y_S (wt%)	Thickness of Film, L (cm)	ΔL (cm)	Bulk Resistance, R_b (Ω)	ΔR_b (Ω)	DC Conductivity, σ_{DC} (S cm^{-1})	$\Delta \sigma_{DC}$ (S cm^{-1})
12.0	0.020^a	5.77 × 10^{-4}	7.41 × 10^{4a}	4.03 × 10^3	8.48 × 10^{-7a}	4.28 × 10^{-8}
	0.021^a		1.92 × 10^4		3.27 × 10^{-6}	
	0.020^a		6.58 × 10^4		9.55 × 10^{-7}	
			7.98 × 10^{4a}		7.87 × 10^{-7a}	
Mean	0.020		7.70 × 10^4		8.18 × 10^{-7}	

^{a}Selected value used for the calculation of mean and standard deviations.

TABLE 5.2 The Standard Deviations ΔL, ΔR_b, and $\Delta \sigma_{DC}$ for PEO/PMMA 25/75 Doped with 12 wt% LiClO$_4$ Calculated after eq 5.8 Where the R_b Value Is Estimated Using the Graphical Method.

Mass Ratio of LiClO$_4$, Y_S (wt%)	Thickness of Film, L (cm)	$\Delta L(c)$	Bulk Resistance, R_b (Ω)	ΔR_b (Ω)	DC Conductivity, σ_{DC} (S cm^{-1})	$\Delta \sigma_{DC}$ (S cm^{-1})
12.0	0.020^a	5.77 × 10^{-4}	6.63 × 10^{4a}	6.36 × 10^3	9.48 × 10^{-7a}	8.01 × 10^{-8}
	0.021^a		1.92 × 10^4		3.27 × 10^{-6}	
	0.020^a		5.87 × 10^4		1.07 × 10^{-6}	
			7.53 × 10^{4a}		8.35 × 10^{-7a}	
Mean	0.020		7.08 × 10^4		8.92 × 10^{-7}	

^{a}Selected value used for the calculation of mean and standard deviations.

TABLE 5.3 The Mean σ_{DC} Values Are Evaluated after eq 5.7 for PEO/PMMA 25/75 Doped with Different Concentrations of LiClO$_4$ by Applying R_b Values Determined Using the Mathematical and Graphical Methods, Respectively. The Percentage Differences in σ_{DC} Value between the Two Methods are Shown.

Mass ratio of LiClO$_4$, Y_s (wt%)	Mathematical Method			Graphical Method			% Difference in σ_{DC} Value between the Two Methods
	Mean L (cm)	Mean R_b (Ω)	Mean σ_{DC} (S cm^{-1})	Mean L (cm)	Mean R_b (Ω)	Mean σ_{DC} (S cm^{-1})	
7.0	0.031	$2.47 \times 10^5 \pm 2.97 \times 10^4$	$3.97 \times 10^{-7} \pm 4.78 \times 10^{-8}$	0.031	$2.47 \times 10^5 \pm 4.10 \times 10^4$	$4.00 \times 10^{-7} \pm 6.64 \times 10^{-8}$	1
10.0	0.031	$2.75 \times 10^5 \pm 8.91 \times 10^4$	$3.74 \times 10^{-7} \pm 1.21 \times 10^{-7}$	0.031	$2.97 \times 10^5 \pm 1.26 \times 10^5$	$3.60 \times 10^{-7} \pm 1.53 \times 10^{-7}$	4
12.0	0.020	$7.70 \times 10^4 \pm 4.03 \times 10^3$	$8.18 \times 10^{-7} \pm 4.28 \times 10^{-8}$	0.020	$7.08 \times 10^4 \pm 6.36 \times 10^3$	$8.91 \times 10^{-7} \pm 8.01 \times 10^{-8}$	9
15.0	0.031	$5.17 \times 10^5 \pm 9.83 \times 10^4$	$1.92 \times 10^{-7} \pm 3.65 \times 10^{-8}$	0.031	$4.39 \times 10^5 \pm 1.25 \times 10^5$	$2.32 \times 10^{-7} \pm 6.61 \times 10^{-8}$	21
20.0	0.023	$1.64 \times 10^6 \pm 3.54 \times 10^4$	$4.42 \times 10^{-8} \pm 9.56 \times 10^{-10}$	0.023	$1.69 \times 10^6 \pm 8.49 \times 10^4$	$4.28 \times 10^{-8} \pm 2.15 \times 10^{-9}$	3

5.3.2.4 INSERTING THE SEMICIRCLE IN THE NYQUIST PLOT USING PARAMETERS EVALUATED BY THE MATHEMATICAL METHOD

When SPE exhibits a Nyquist plot with a perfect semicircle at the high frequency region, it has an equivalent circuit or a RC circuit comprising of a pure resistor in parallel with a pure capacitor. The equation of the semicircle with radius $R/2$ and centered at $(R/2, 0)$ is given by eq 5.9

$$(Z')^2 + (Z'')^2 = (\frac{R}{2})^2. \qquad (5.9)$$

Generally, most polymer–salt systems do not obey the Debye response when subjected to an applied field, thus, exhibit Nyquist plots with either distorted or depressed semicircles with the center slightly below the x-axis, followed by an inclined spike at the lower frequency region of the plot. According to Winie and Arof,[3] the deviation is due to the fact that polymer electrolyte is a constant phase element (CPE), not a pure capacitor and that a CPE is considered to be a leaky capacitor.

The semicircle represents the bulk resistance of the polymer-salt sample while the inclined spike deals with the interfacial impedance. Therefore, the following discussion concentrates only on procedures to insert the semicircle into the Nyquist plot.

Figure 5.20 depicts a depressed semicircle with the center shifted a distance Δ below the x-axis for PEO doped with 5 wt% of $LiClO_4$. For a

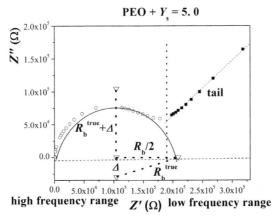

FIGURE 5.20 Illustration of a depressed semicircle of a Nyquist plot.

perfect semicircle centered at $(R_b/2, 0)$, $Z' = Z''$, so for the depressed semi-circle, the value of Δ is the difference between Z' and Z''. The values of f_o, Z', Z''_{max}, and R_b for the polymer electrolyte, PEO/PMMA 75/25 doped with 10 wt% of LiClO$_4$ determined in Section 5.3.2.2 are to be applied in this section to insert the semicircle into the Nyquist plot of this sample. The following values are obtained from Section 5.3.2.

$$f_o = 50000 \text{ s}^{-1}$$

$$Z' = 1.82 \times 10^{-4} \ \Omega$$

$$Z'' = 1.54 \times 10^{-4} \Omega$$

$$\Delta = (1.82 \times 10^{-4} - 1.54 \times 10^{-4})\Omega$$

$$= 2800\Omega$$

$$R_b = 3.64 \times 10^{-4} \text{ S cm}^{-1}.$$

The experimental data of the Nyquist plot obey eq 5.10:

$$|Z|^2 = (Z')^2 + (Z'')^2$$

$$\text{in which } |Z| = \frac{Z_o}{2} \equiv \frac{R_b}{2}. \tag{5.10}$$

The coordinates of the center of the depressed semicircle (below x-axis) are given as:

$$Z'_{(centre)} = \frac{R_b}{2} \text{ and } Z''_{(centre)} = -\Delta. \tag{5.11}$$

Inserting eq 5.11 into eq 5.10 gives eq 5.12 as follows:

$$\left(Z' - \frac{R_b}{2}\right)^2 + (Z'' + \Delta)^2 = \left(\frac{R_b}{2}\right)^2 \tag{5.12}$$

$$0.83\left[1 + \cos(n\pi/2)\right] - \sin(n\pi/2) = 0. \tag{5.13}$$

Substituting the values of Δ and R_b into eq 5.13 and changing Z' and Z'' into x and y, respectively, gives eq 5.14. By plotting eq 5.14, a semicircle is inserted into the Nyquist plot of the sample as shown in Figure 5.21:

Impedance Spectroscopy: A Practical Guide

$$y = -2.8 \times 10^3 + \sqrt{(3.64 \times 10^4 x - x^2)}. \qquad (5.14)$$

Or in computer language,

$$y = -2.8e^3 + (3.64e^4 * x - x * x)^{\wedge}(0.5).$$

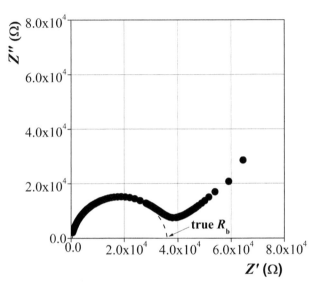

FIGURE 5.21 Insertion of a semicircle into a Nyquist plot of PEO/PMMA 75/25 doped with 10 wt% $LiClO_4$.

5.4 ESTIMATION OF DEBYE RELAXATION EXPONENT, N

The graphical and the mathematical methods to determine the R_b values for SPE samples have been discussed in detail in Section 5.3. Apart from the value of R_b, the frequency f_o and the Z' value at which the Z'' curve reaches its peak with the value of Z''_{max} can also be attained as described in Section 5.3.2.2.

Debye's relaxation approach is an empirical approach dealing with the dielectric response of ideal, perfect, or non-interacting assembly of dipoles to an alternating electric field. Materials such as semiconductors, ceramics, etc. that obey the Debye relaxation model possesses only a single relaxation time because the dipoles which neither interact with nor induce other dipoles surrounding them, are able to simultaneously align themselves along the direction of the applied field. The Debye relaxation model is expressed in eq 5.15

$$Z = \frac{Z_o}{1+(i\omega\tau)}$$

τ is the relaxation time which is the time taken for an ionic charge carrier to align with the applied field.

ω is the angular frequency in which $\omega = 2\pi f$

Since $Z_o = R_b$

$$\therefore Z = \frac{R_b}{1+(i\omega\tau)}$$

$$Z = \frac{R_b(1-i\omega\tau)}{[1+(i\omega\tau)](1-i\omega\tau)}$$

$$Z = \frac{R_b(1-i\omega\tau)}{1+(\omega\tau)^2}$$

$$Z = \frac{R_b}{1+(\omega\tau)^2} - i\frac{R_b(\omega\tau)}{1+(\omega\tau)^2}$$

As $(Z = Z' + iZ'')$

$$\therefore Z' = \frac{R_b}{1+(\omega\tau)^2} \quad \text{and} \quad Z'' = \frac{R_b(\omega\tau)}{1+(\omega\tau)^2}.$$

$$(5.15)$$

Impedance Spectroscopy: A Practical Guide

The phase angle ($\omega\tau$) in radian or (θ) in degree represents the lag between the AC voltage and its current. For ideal case (obey Debye model) $\theta = 90°$ and $\omega\tau = 1$ implying that the material has a purely capacitive behavior. Meanwhile $\theta = 0°$ and $\omega\tau = 0$ indicates that it has purely resistive behavior. However, most materials especially polymeric materials commonly show deviation from the ideal Debye behavior. Therefore, some generalization has been made to the Debye relaxation equation in order to accommodate materials which do not obey this relationship. The modified Debye equation is given in eq 5.17.

$$Z(\omega) = \frac{Z_o}{1+(i\omega\tau)^n}$$

$$= \frac{Z_o\left[1-(i\omega\tau)^n\right]}{\left[1+(i\omega\tau)^n\right]\left[1-(i\omega\tau)^n\right]}$$

$$= \frac{Z_o\left[1-(i\omega\tau)^n\right]}{1+(\omega\tau)^{2n}}$$

$$= \frac{Z_o}{1+(\omega\tau)^{2n}} - i\frac{Z_o(\omega\tau)^n}{1+(\omega\tau)^{2n}} \tag{5.16}$$

$$\therefore Z'(\omega) = \frac{Z_o}{1+(\omega\tau)^{2n}}, \qquad Z'' = \frac{Z_o(\omega\tau)^n}{1+(\omega\tau)^{2n}}. \tag{5.17}$$

The Debye exponent "n" takes values between zero and one, that is, $0 < n < 1$. Thereby, if $n = 1$, the material is a perfect capacitor (Debye's relaxation) whereas if $n = 0$, it is a perfect resistor. Nevertheless, the Debye exponent for most polymeric materials falls in the range of 1–0.8 which is the most acceptable range.

The following procedures are presented to evaluate the Debye's relaxation exponent, n of SPE. It is worthwhile to note that this method only applied to SPE samples with relatively good ionic conductivity in which both the frequency dependent plots of Z' and Z'' intersect as shown in

Figure 5.11(b) in Section 5.3.2.1. In general, the relationship between the values of Z', Z", and the phase angle, θ are shown in Figure 5.22.

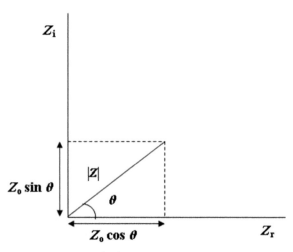

FIGURE 5.22 The magnitude of $|Z|$, Z', Z" and the phase angle, θ of an impedance Z.

The magnitude of $|Z|$ is defined in eq 5.10 in Section 5.3.2.4. The expressions $Z_o \cos \theta$ and $Z_o \sin \theta$ represent the real and imaginary part of the impedance, respectively. The procedure starts with the Debye equation after generalization which is eq 5.17. After going through some mathematical operations, the magnitudes of Z' and Z" are expressed in terms of cos $(n/2)$ and sin $(n/2)$ as given in eq 5.18

$$Z(\omega) = \frac{R}{1+(i\omega\tau)^n}$$

$$Z_r(f) = R\frac{1+(\omega\tau)^n \cos\left(n\frac{\pi}{2}\right)}{N}$$

$$Z_i(f) = R\frac{(\omega\tau)^n \sin\left(n\frac{\pi}{2}\right)}{N}$$

where $N = 1 + 2(\omega\tau)^n \cos\left(n\frac{\pi}{2}\right) + (\omega\tau)^2.$ (5.18)

Impedance Spectroscopy: A Practical Guide 125

As mention in Section 5.3.2.1, the Z'' curve reaches its peak or maximum value at f_o, therefore the $Z''(f)$ in eq 5.18 becomes $Z''_{max}(f_o)$. To evaluate the value of n, eq 5.19 is generated by taking the ratio of $Z''_{max}(f_o)/Z'(f_o)$ from eq 5.18.

$$\frac{Z''_{max}(f_o)}{Z'(f_o)} = \frac{\sin\left(n\frac{\pi}{2}\right)}{1+\cos\left(n\frac{\pi}{2}\right)}.$$

$$(5.19)$$

The values of f_o, Z''_{max}, and Z' can be determined by referring to the detail procedures described in Section 5.3.2.2. For example, if the values of Z''_{max} and Z' obtained for this sample are: $Z''_{max} = 2.62 \times 10^4\ \Omega$; $Z' = 3.14 \times 10^4\ \Omega$, and substituting these values into eq 5.19, the magnitude of the ratio of $Z''_{max}(f_o)/Z'(f_o)$ is equal to 0.83 as illustrated below

Equation 5.19 can now be $$\frac{Z''_{max}(f_o)}{Z'(f_o)} = \frac{2.62\times10^4\Omega}{3.14\times10^4\Omega}\ \text{written as}$$

$$= 0.83.$$

$$0.83\left[1+\cos(n\pi/2)\right]-\sin(n\pi/2)=0. \qquad (5.20)$$

Or in computer language:

$$0.83*\left(1+\cos(1.57*x)\right)-\sin(1.57*x)$$

To evaluate the n value, a sinusoidal graph based on eq 5.19 is constructed as shown in Figure 5.23(a) and later obtained the precise value of n as shown in Figure 5.23(b) by referring to Section 5.3.2.2 on the procedures for inserting a function graph.

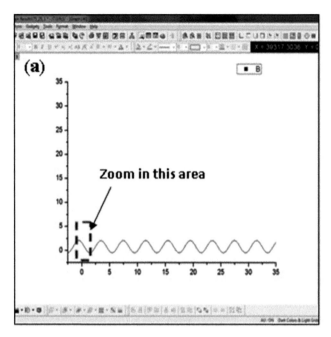

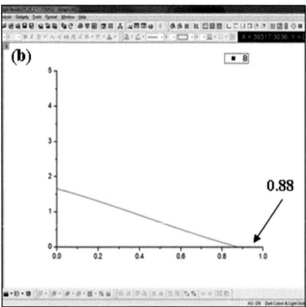

FIGURE 5.23 (a) a sinusoidal graph based on eq 5.19, (b) adjusting the scales of the x- and y-axes to determine the exponent n value.

Impedance Spectroscopy: A Practical Guide

5.5 SUMMARY

In this chapter, practical guides to evaluate parameters of Nyquist plot using two different methods are discussed. The chapter begins with the introduction of different forms of Nyquist plots and the relationship between the real and imaginary components of complex impedance. Two methods, the graphical and the mathematical method, are applied to determine the R_b and σ value of SPE. Every guided procedure carry out in the mathematical method to determine the values of R_b, σ_{DC}, f_o, Z', Z'', and the relaxation exponent n is thoroughly explained with examples and illustrations. Standard operation procedures in using the Origin® software to perform polynomial curve fitting and to insert functional graphs are presented in detail.

5.6 ACKNOWLEDGMENTS

The authors would like to express our gratitude toward Ministry of Higher Education (MOHE) for the "Fundamental Research Grant Scheme" grant (FRGS) (600-RMI/FRGS 5/3(67/2013)) supporting the research work.

KEYWORDS

- **solid polymer electrolyte**
- **impedance spectroscopy**
- **bulk resistance**
- **ionic conductivity**

REFERENCES

1. Chandra, S. *Superionic Solids: Principles and Applications;* Amsterdam: North Holland, 1981.
2. MacDonald, J. R. Impedance Spectroscopy: Old Problems and New Developments. *Electrochim. Acta.* **1990,** *35,* 1483–1492.
3. Winie, T; Arof, A. K. Impedance Spectroscopy: Basic Concepts and Application for Electrical Evaluation of Polymer Electrolytes. In *Physical Chemistry of*

Macromolecule - Marco to Nanoscales; Apple Academic Press: Toronodo, NJ, 2013; pp 500–570.

4. Boukamp, B. A. A Package for Impedance/Admittance Data Analysis. *Solid State Ion.* **1986a,** *18–19,* 136–140.
5. Boukamp, B. A. A Nonlinear Least Squares Fit procedure for Analysis of Immittance Data of Electrochemical Systems. *Solid State Ion.* **1986b,** *20* (1), 31–44.
6. Bruce, P. G. Structure and Electrochemistry of Polymer Electrolytes. *Electrochim. Acta.* **1995,** *40,* 2077–2085.
7. Subban, R. H. Y.; Mohamed, N. S.; Arof, A. K. *Electrochemical Power Sources – Materials and Characterization;* University Malaya Publisher: Kuala Lumpur, 2000; pp 15–29.
8. Baskaran, R.; Selvasekarapandian, S.; Kuwata, S.; Kawamura, J.; Hattori, T. Conductivity and Thermal Studies of Blend Polymer Electrolytes Based on PVAc–PMMA. *Solid State Ion.* **2006,** *177,* 2679–2682.

CHAPTER 6

POLYMERIZATION OF NATURAL OILS FOR A QUARTZ CRYSTAL MICROBALANCE-BASED GAS SENSOR APPLICATION

RASHMITA DAS[1,*], PANCHANAN PRAMANIK[2], and RAJIB BANDYOPADHYAY[1]

[1]*Department of Instrumentation and Electronics Engineering, Jadavpur University, Kolkata 700098, West Bengal, India*

[2]*Department of Basic Science, MCKV Institute of Engineering, Liluah, Howrah 711204, West Bengal, India*

Corresponding author. E-mail: kunurashmita@gmail.com

CONTENTS

Abstract 130
6.1   Introduction 130
6.2   Cross-Linked Polymers from Oils 134
6.3   Applications of Polymerized Oil 135
6.4   Sensor 137
6.5   Case Study 1: Polymerized Tung Oil-Based Quartz Crystal Microbalance Sensor for the Detection of Amines Vapors 144
6.6   Case Study 2: Polymerized Linseed Oil Based Quartz Crystal Microbalance Sensor and Its Application 152
6.7   Conclusion 157
Keywords 157
References 158

ABSTRACT

Recently, various types of useful condensation polymers, such as polyurethanes, polyesters and polyethers, are being produced from natural oils. Direct polymerization of oils has been used in coatings industries for a long time. It is also found that, these oils are used in food industries as well as for the production of coatings, inks, plasticizers, lubricants, and agrochemicals. Oils can be thermally polymerized by heating at high temperature and the products are oligomers of high viscosity. In this chapter, we have discussed about the polymerization of natural oils by different methods and the applications in the detection of volatile organic compounds. Free radical polymerizations of tung oil, linseed oil, and castor oil are carried out by using benzoyl peroxide as initiator. Quartz crystal microbalance (QCM) is a mass-sensitive device. Its principle is based on the interactions between the coated surface of quartz crystal and the analytes. The developed sensors are used for the detection of volatile organic compounds like benzene, toluene, o-xylene, p-cresol, and aliphatic amines. It is observed that the polymerized linseed oil-based sensors show good sensitivity and selectivity toward benzene, toluene, o-xylene, and p-cresol whereas the polymerized castor oil and polymerized tung oil-based sensors are useful for aliphatic amines detection.

6.1 INTRODUCTION

Natural oils are the most important renewable sources in the preparation of various industrial materials. These are generally obtained from the naturally occurring plants like sunflower, linseed, cotton etc. Triglycerides are the main constituents of the plant oils which are formed by the esterification of glycerol with three fatty acids. Triglycerides chains having internal double bonds, alcohols, or epoxides, can be polymerized by different methods, including condensation, radical, cationic, and metathesis procedures. Polymers are generally obtained from natural or synthetic sources. These are generally made up of long covalently bonded molecules. It is reported that, plant oils are used as a natural renewable resource for the production of polymeric materials.[1] Plant oils have many advantages like world-wide availability and relatively low prices which make them industrially attractive and feasible. A minor modification reaction of these natural oils has to be performed in order to obtain suitable monomers for many different applications. The physical and chemical properties of such oils are affected by the stereochemistry of the double bonds of the fatty acid chains, their degree of unsaturation, and also the length of the carbon chain. The syntheses of cross-linked polymers from triglycerides are possible due to its highly functionalized molecules. The functional groups present in triglycerides, such as internal double bonds, alcohols, or epoxides can be polymerized easily by different methods. The chemical structures of some triglycerides oils are given in Table 6.1. The chemical modifications of natural oils are also possible prior to polymerization of triglycerides chain. By this process, we can introduce polymerizable functional groups to natural triglycerides chain having only double bond, which enhanced the synthetic possibilities.[2]

TABLE 6.1 Chemical Structure of Some Triglycerides Oils.

Triglycerides Oil	Chemical Structure
Linseed oil	
Castor oil	
Sunflower oil	
Soybean oil	

The total weight of one molecule of triglyceride oil contain 94–96% of fatty acids, however, it has been known that some of the fatty acids are saturated and some are unsaturated acids having conjugated double bonds or may have isolated double bond. Natural oils also have fatty acid chains having hydroxyl, epoxy or Oxo groups, or triple bonds on its backbone. Hence, the physical properties are different due to the different structures of fatty acids of the natural oils. Common oils such as linseed, sunflower, castor, soybean, palm, and rapeseed oils are used for the synthesis of polymer. Since castor oil contains ricinoleic acid ((9Z,12R)-12-hydroxy-9-octadecenoic acid), it provides additional natural chemical functionality for modifications, cross-linking, or polymerization than linseed oil. The fatty acid contents in various natural oils are tabulated in Table 6.2.[3]

TABLE 6.2 Major Fatty Acid Composition in Natural Oils.

Fatty Acid	Castor Oil (%)	Linseed Oil (%)	Cotton Seed Oil (%)	Soybean Oil (%)	Tung Oil (%)	Sunflower Oil (%)
Oleic acid	5	19.1	18.6	23.4	8	42
Stearic acid	0.5	3.5	2.6	4.0	4	4
Palmitic acid	1.5	5.5	21.6	11.0	–	6
Linolenic acid	0.5	56.6	0.7	7.8	–	47
Linoleic acid	4	15.3	54.4	53.3	4	1
Alpha-eleostearic acid	–	–	–	–	84	–

Natural oils are less toxic, easily available, relatively low price and biodegradable.[4] Plant oils are commonly used in the paint formulation. In the last few decades, oil-based polymers have been focused for different industrial applications such as surfactants, coatings, lubricants, adhesives, drying agents, cosmetics, printing inks, emulsifiers, and plasticizers. Some of the polymers prepared from triglycerides are ox-polymerized oils, polyesters, polyurethanes (urethane oils), polyamides, acrylic resins, epoxy resins, and polyester amides etc.[3] The details of some triglycerides oils are described below.

6.1.1 CASTOR OIL

This oil is obtained from the seed of *Ricinus communis*. It is light yellow in color, has high specific gravity, high viscosity, and solubility in alcohol.

This oil is used for the preparation of resins and plasticizer in its unhydrated form. This oil contains a hydroxyl group, which is the basic difference from other oils. Chemical dehydration may take place between the hydroxyl group and the adjacent hydrogen atom by the removal of water molecules to convert it to unsaturated fatty acid ester. It is also known that, dehydrated castor oil is used as drying oil in paint industries.

6.1.2 LINSEED OIL

Linseed oil is highly unsaturated oil, which is obtained from the seed and widely used as the drying oil in paint industries and as an ingredient for the modification of resins. This oil is obtained by solvent extraction method which contains low impurities and better quality.

6.1.3 COTTON SEED OIL

This oil is used for the manufacturing of alkyd resins and also used in drying oil in paints. This oil is obtained from the plant *Gossypium malvaceae*.

6.1.4 SOYBEAN OIL

This oil is commercially important for the use in interior and exterior paints and also in the preparation of alkyds resins. This oil is obtained from the plant *Soja hispida*.

6.1.5 TUNG OIL

This is the Chinese wood oil obtained from *Aleurites fordii* and *A. montana*. This oil is extensively used for exterior varnish and alkyd resin preparation and also used as a water-resistant coating in vehicles. It contains conjugated double bonds, which can be easily polymerized under heat treatment. It has a high specific gravity, high viscosity, and high refractive index.

6.1.6 SUNFLOWER OIL

This oil is obtained from *Helianthus annuus*. It is commonly used as semi-drying oil, commercially important in coating industries.

6.2 CROSS-LINKED POLYMERS FROM OILS

Linseed oil is polymerized by cationic, thermal, free radical process, as well as by the oxidative polymerization method.[5] This drying oil is used for surface-coating applications. Copolymerization of linseed oil can be achieved by using different olefinic monomers.[6–8] Cobalt, lead, and zirconium-2-ethylhexanoates are generally used as catalysts for the oxidative polymerization of linseed oil. Linseed oil has been styrenated for using in the polymerizations process to obtain some desired film properties.[9] Free radicals are formed by thermolysis, and the formation of free radicals depends upon the degree of conjugation and unsaturation on oil. The macro monomers of linseed oil were prepared by transesterification of methylmethacrylate (MMA) with partial glycerides.[10] The schematic diagram of polymerization of linseed oil is given in Figure 6.1. Auto-oxidation of linseed oil is also possible through the hydrocarbon oxidation by the reactive allylic hydrogen atoms of the chain.[11,12] A combination of any radical formed in the system or by their direct addition to double bonds leads to the formation of three-dimensional cross-linked structures.

Castor oil has been used for the preparation of interpenetrating polymer networks (IPN) which shows the dynamic mechanical behavior.[13] The hydroxyl groups of castor oil form hydrogen bonds with other polymer matrix. Styrenated castor oil and linseed oil by the macromer technique has been reported earlier.[14] Also, the copolymerization of dehydrated castor oil with styrene is already been established.[15] It is observed that when the concentration of dehydrated castor oil is more than 20%, the copolymerization is quite difficult. The synthesis of IPNs with cross-linked polystyrene and castor oil elastomers has also been reported previously.[16–18]

Tung oil is extracted from the seeds of the tung tree. The main components are glyceride of elaeostearic acid with a conjugated triene structure as described above. Tung oil undergoes rapid polymerization by both free radical and cationic polymerizations due to highly unsaturated and conjugated system.[19,20] Tung oil is copolymerized with divinylbenzene to form hard plastics in the presence of boron trifluoride diethyl etherate.[21] The

Polymerization of Natural Oils   135

polymeric solid is formed within a few seconds after the addition of this initiator at room temperature. The polymerized tung oil is thermally stable up to 200°C. Also the incorporation of aromatic comonomers produces viable polymeric materials from elastomers to tough and rigid plastics. Highly cross-linked bulk polymer is formed from tung oil with the addition of metallic catalysts. The copolymerization of tung oil with diacrylate by using 1,6-hexanediol diacrylate and 1,4-butanediol diacrylate by Diels–Alder reaction is also reported and the copolymer exhibit good solvent resistance, high hardness, and good gloss.[22]

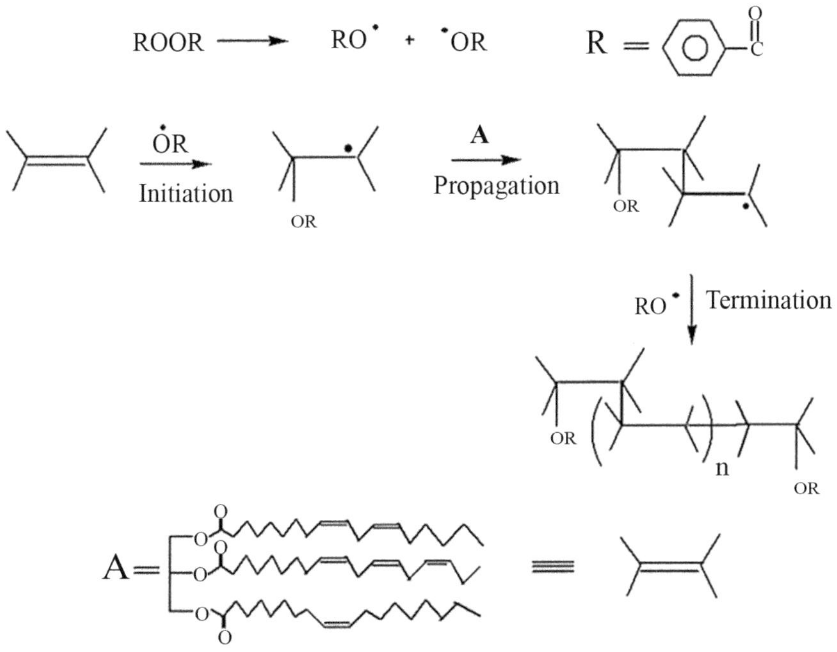

FIGURE 6.1   Reaction mechanism of polymerization of linseed oil (Ref. [78]).

6.3   APPLICATIONS OF POLYMERIZED OIL

Although the biggest usable area of this polymerized oil is in coating industry, in the last few decades, triglyceride oil-based polymers have been used for many different applications. Linseed, sunflower, rapeseed oils, and soybean oil-based resins are commonly used in the printing industry. Polyesters, polyurethanes (urethane oils), polyamides, acrylic resins, epoxy resins, and polyester amides are prepared from triglyceride oils. The cross-linked polymers have high modulus, high fracture strength, and solvent resistance. It is also found that highly cross-linked polymers are used for synthesis of composites, foamed structures, structural adhesives, insulators for electronic packaging, and other applications.[23] Without using any activator, soybean oil can be oxidized by permanganate oxidation with sub/supercritical CO_2. The oxidized oil is used as semi-drying oil in paint industries.[24]

The esterification of polyhydroxy alcohols with polybasic acids and fatty acids gives alkyd resin polymer which can also be produced from triglyceride oils. Alkyd resins have wide applications due to its cost-effective and biodegradable properties. The alkyd resins prepared from 40% oils are short oil alkyds, and are used for baked finishes on automobiles, refrigerators, stoves, washing machines etc. Above 60% oil, alkyd resins are used for the synthesis of long alkyd oils which are used in brushing enamels. It is observed that, oil-based polyhydroxyalkanoates (PHAs) are alternative polyesters[25–29] in paint formulation. They are water-insoluble polyesters of carbon, oxygen, hydrogen, and are optically active, biodegradable in nature. The crystallinity of the polymer decreases by the incorporation of a small amount of long-chain monomers, but the toughness and flexibility of the synthesized polymer will be increased. It has been observed that, for all the above purposes, olive, castor,[30] tallow,[31,32] soybean, sunflower (high oleic), coconut oils[33] and soybean,[34] linseed and tall oil fatty acids[28,35] are used as monomers.

Poly (propylene glycol) and castor oil blends are used for the preparation of millable polyurethane elastomers which can be used for the industrial applications.[36] Castor oil can also be used for the preparation of polyurethane network by the reaction with diisocyanate. The polymers prepared from the trienoic acid have good film properties such as short drying time, good water, alkali, and acid resistances. The concentrations of diisocyanates used for the preparation of polymer film affect its mechanical and gas permeability properties. It has also been observed that seed oils of *Ecballium elaterium* and *P. mahaleb* are used for the preparation of oil-modified polyurethanes.[37]

Nylon 11 is a commercially used polymer, which has been prepared from castor oil. It has a wide range of flexibility, excellent dimensional stability and electrical properties, good chemical resistance, and low cold brittleness temperature. Although vinyl modified triglyceride oils are mainly used in paint industry, in recent years, there has been increasing trend toward their use as biopolymers.[37,38] Grafted copolymers can be prepared from linseed, soybean oils, methyl methacrylate, styrene, or n-butyl methacrylate by the conversion of oil to polymeric peroxide under atmospheric conditions or O_2 gas at room temperature. These synthesized polymers are used for biomedical applications.[39,40] Epoxidized seed oils are widely used for the synthesis of cationic UV-curable coatings.[41–49] The adhesion, toughness, water, and chemical resistance properties are enhanced by using linseed oil-based polyester amide modified with toluene diisocyanate. Epoxidized oils are found to be good plasticizers as well as stabilizers for poly vinyl chloride. Poly(methyl methacrylate)-based graft copolymers are also useful for biomedical applications.[50,51] Soybean oil-based polyester amide urethanes filled with boron shows combined properties of polyester and polyamide as well as anti-microbial properties.[52] Urushi is a traditional lacquer used in art and household materials in Japan. This polymer is prepared from triglyceride oils. From literature survey, it is found that the electrode material is prepared from a composite of graphite and polyurethane resin; it can also obtained from castor oil.[53] Epoxidized soybean oil-based composite exhibited strong viscoelastic solid properties similar to synthetic rubbers. Soybean oil-based resin and natural fibers can be used as a material for roof, floors, or walls of a house or low-rise commercial buildings.[54,55]

6.4 SENSOR

A sensor is a device that detects changes in quantities and provides a corresponding electrical or optical signal. A common example is thermocouple, which converts temperature to an output voltage and mercury-in-glass thermometer, it can measure temperature as the expansion and contraction of a liquid which can be read on a calibrated glass tube. The common features to be considered to choose sensor are accuracy, environmental condition, range, calibration, resolution, cost, and repeatability. In many industries, the quality estimation of food, drinks, perfumes, cosmetic, and chemical products are carried out by the human nose through its color and odor. These processes are not portable, they tend to be expensive, and their performance is relatively slow. Hence nowadays, identification and discrimination of products are carried out by utilizing an electronic nose (e-nose) system which is basically an array of sensors to give a fingerprint response to a given odor, and pattern recognition software.

Classification of sensors based on the properties is given below

- Temperature—Thermistors, thermocouples, RTD's, IC and many more.
- Pressure—Fiber optic, vacuum, elastic liquid-based manometers, LVDT, electronic.
- Flow—Electromagnetic, differential pressure, positional displacement, thermal mass, etc.
- Level sensors—Differential pressure, ultrasonic radio frequency, radar, thermal displacement, etc.
- Proximity and displacement—LVDT, photoelectric, capacitive, magnetic, ultrasonic.
- Biosensors—Resonant mirror, electrochemical, surface plasmon resonance, light addressable potentiometric.
- Image—Charge-coupled devices, CMOS.
- Gas and chemical—Semiconductor, infrared, conductance, electrochemical.
- Acceleration—Gyroscopes, accelerometers.

Others sensors are moisture, humidity sensor, speed sensor, mass, tilt sensor, force, and viscosity. We have discussed about a gas sensor application using quartz crystal microbalance.

6.4.1 GAS SENSOR

A gas sensor is a transducer that detects gas molecules and produces an electrical signal with a magnitude proportional to the concentration of the gas. The types of measurement deals with voltage, temperature, humidity and the measurement of gases are much more complicated. Because there are literally hundreds of different gases, and there is a wide array of diverse applications in which these gases are present, each application must implement a unique set of requirements. For example, some applications may require the detection of one specific gas, while eliminating readings from other background gases. Conversely, other applications may require a quantitative value of the concentration of every gas present in the area.

6.4.2 APPLICATION OF GAS SENSORS

All of the sensors are used for a wide range of applications, few are given below.

- Fire/smoke detection
- Gas boiler safety switch
- CO, NH_3, H_2S alarm
- Inflammable gas detection
- Air purifier
- Air conditioner
- Cabin air quality
- LPG alarm for RV
- Microwave oven
- Parking garage ventilation control
- Breath alcohol tester
- Gas-driven car
- Ventilation control etc.

Generally, sensors used for electronic nose systems are metal oxide sensors, surface acoustic wave sensor, quartz crystal microbalance sensor, etc. We are interested for quartz crystal microbalance sensor (QCM) for our research work.

6.4.3 METAL OXIDE SENSORS

Metal oxide sensors are also known as chemiresistors. The detection principle of resistive sensor is based on the change of the resistance of a thin film upon adsorption of the gas molecules on the surface of a semiconductor. The gas–solid interactions affect the resistance of the film because of the density of electronic species in the film.

6.4.4 SURFACE ACOUSTIC WAVE SENSOR

This is a class of microelectromechanical systems (MEMS) which rely on the modulation of surface acoustic waves to sense a physical phenomenon. This type of sensor transduces an input electrical signal into a mechanical wave then to electrical signal. It measures the changes in amplitude, phase, frequency, or time-delay between the input and output electrical signals. These types of sensors are useful for chemical vapors, biological matter, humidity, ultraviolet radiation, magnetic fields of magnetic materials, and viscosity of a liquid phase measurement.

6.4.5 QUARTZ CRYSTAL MICROBALANCE SENSOR

Quartz crystal microbalance (QCM) is a device consisting of a thin quartz disk coated with silver electrodes on both sides. The quartz crystal plate must be AT or BT cut with specific orientation with respect to the crystal axes, so that the acoustic wave propagates perpendicularly to the crystal surface. The resonant frequency of the quartz crystal depends on the angles with respect to the optical axis at which the wafer was cut from the crystal. The AT-cut quartz crystal has nearly zero frequency drift with temperature around room temperature, which is a great advantage of it. However, in 1959, Sauerbrey showed that the frequency shift of a quartz crystal resonator is directly proportional to the mass loaded on the surface. When voltage is applied to a quartz crystal causing it to oscillate at a specific frequency, the change in mass on the quartz surface is directly related to the change in frequency of the oscillating crystal, as shown by the Sauerbrey equation (eq 6.1).

$$\Delta f = -\frac{2f_0^2}{A\sqrt{\rho_q \mu_q}} \Delta m \tag{6.1}$$

f_0—resonant frequency (Hz)
Δf—frequency change (Hz)
Δm—mass change (g)
A—piezoelectrically active crystal area (Area between electrodes, cm^2)
ρ_q—density of quartz ($\rho_q = 2.648$ g/cm^3)
μ_q—shear modulus of quartz for AT-cut crystal ($\mu_q = 2.947 \times 10^{11}$ g·cm^{-1}·s^{-2})

The Sauerbrey equation is only strictly applicable to uniform, rigid, thin-film deposits. However, liquid phase QCM measurements have been done from the beginning of 1980s. The oscillating frequency of a piezoelectric crystal decreases with the adsorption of any substances on the surface.

Theoretically, the detection limit of oscillating quartz crystals is about 10^{-12} g. The QCM has large applications for trace amounts of gas detection, immunosensors, DNA biosensors, and drug analysis.[56] The adsorption of antibodies is a method which is widely used in the detection of various analytes,[57–60] which offers a high affinity and specific binding site.

The liquid phase QCM measurements have been done by Kanazawa and coworkers, who showed that the change in resonant frequency of a QCM taken from air into a liquid proportional to the square root of the liquid's density–viscosity product as given in eq 6.2.

$$\Delta f = -f_0^{3/2} (\rho_L \mu_L / \pi \rho_q \mu_q)^{1/2} \tag{6.2}$$

where Δf = measured frequency shift,

f_0 = resonant frequency of the unloaded crystal,

ρ_L = density of liquid in contact with the crystal,

μ_L = viscosity of liquid in contact with the crystal,

ρ_q = density of quartz,

μ_q = shear modulus of quartz.

6.4.6 APPLICATIONS OF QCM

The high sensitivity and the real-time monitoring of mass changes of the crystal make QCM a very attractive technique for a large range of applications. The sensitivity of the QCM is approximately 100 times higher than an electronic fine balance with a sensitivity of 0.1 mg. That means, it is capable of measuring mass changes by loading a monolayer of atoms. QCM systems can be used in fluids or with visco-elastic deposits which has increased the interest toward this technique. Major advantages of the QCM technique used for liquid systems are that it allows a label-free detection of molecules. The applications of QCM in different fields are tabulated in Table 6.3.

In our work, we have used QCM device as a gas sensor for the detection of volatile organic compounds like benzene, toluene, o-xylene, p-cresol, methylamine, ethylamine, tert-butyl amine, etc. Volatile organic compounds (VOCs) are generally a large group of carbon-based chemicals that easily evaporate at room temperature. Although VOCs are found in all living things, but the majority of VOCs is man-made. The volatile organic compounds are important areas of studies for monitoring the status of

TABLE 6.3 Applications of QCM in Different Field.

Electrochemistry	• Thin film thickness monitoring in thermal, e-beam, sputtering, magnetron, ion and laser deposition.
	• Electrochemistry of interfacial processes at electrode surfaces
Biotechnology	• Interactions of DNA and RNA with complementary strands
	• Specific recognition of protein ligands by immobilized receptors, immunological reactions
	• Detection of virus capsids, bacteria, mammalian cells
	• Adhesion of cells, liposomes and proteins
	• Biocompatibility of surfaces
	• Formation and prevention of formation of biofilms
Functionalized Surfaces	• Creation of selective surfaces
	• Lipid membranes
	• Polymer coatings
	• Reactive surfaces
	• Gas sensors
	• Immunosensors
Thin Film Formation	• Langmuir and Langmuir-Blodgett films
	• Self-assembled monolayer
	• Polyelectrolyte adsorption
	• Spin coating
	• Bilayer formation
	• Adsorbed monolayer
Surfactant Research	• Surfactant interactions with surfaces
	• Effectiveness of surfactants
Drug Research	• Dissolution of polymer coatings
	• Molecular interaction of drugs
	• Cell response to pharmacological substances
	• Drug delivery

pollutions in the environment, because these pollutants enter water, soil, and atmosphere are very much harmful for the human health.[61–64] It is due to the carcinogenic, teratogenetic, or mutagenic behavior of VOCs.[65–67] Breathing low concentrations of VOCs for long periods of time may increase some people's risk of health problems. The extent and nature of the health effect will depend on many factors including level of exposure and length of time exposed etc. Some people have experienced eye and respiratory tract irritation, headaches, dizziness, visual disorders, and memory impairment after exposure to some organic vapors. However, it is also known that the exhaled human breath contains a few hundred of volatile organic compounds and is used in breath analysis to serve as a VOC biomarker to test for diseases such as lung cancer, diabetics, liver disorder etc.[68]

We are using a laboratory-made steady state system for frequency measurement of QCM sensor. A schematic diagram of experimental set up is given in Figure 6.2. It consists of sensor chamber, vacuum pump, QCM, PC with Lab view software, National instrumentation card, 8284B IC, resistors, and capacitor. QCM surfaces are modified by a polymer coating before carrying out the experiment. Then the prepared sensor is placed in the sensor chamber and the frequency is measured through a multi-channel frequency counter in a personal computer.

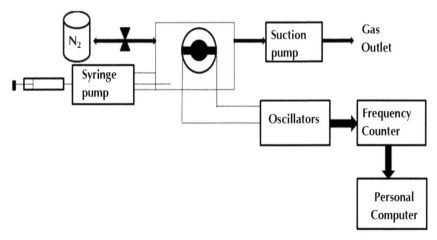

FIGURE 6.2 Schematic diagram of experimental setup.

6.5 CASE STUDY 1: POLYMERIZED TUNG OIL-BASED QUARTZ CRYSTAL MICROBALANCE SENSOR FOR THE DETECTION OF AMINES VAPORS

6.5.1 INTRODUCTION

The sensitivity of QCM sensor is determined by the chemical structure and physical properties of polymeric coating and also by the nature of interaction between polymer coating and vapor molecules. In this work, we have tried to develop polymerize tung oil-based quartz crystal microbalance (QCM) sensors for the detection of organic vapors such as methylamine, ethylamine, diethylamine, triethylamine, and tert-butylamine. The polymer film on QCM surface is prepared by the free radical polymerization of tung oil in chloroform solvent by using benzoyl peroxide as initiator. The sensors are fabricated by solution dip–dry method followed by heating at 110°C under argon atmosphere to get cross-linked film on QCM surface. Six sensors are made by varying the concentration of benzoyl peroxide in combination of tung oils. The responses of the sensors for the concentration range from 5 to 250 ppm of all the amine vapors are observed to be linearly correlated with the concentrations of amine vapors. The sensors show good reproducibility and reversibility. The structures of the polymeric films are confirmed by FTIR analysis. The surface morphologies of the films before and after absorption have been investigated by AFM. The plant oils are found to be an alternative resource for the production of polymeric materials. These oils are relatively low cost and easily available worldwide.[69,70] Larger varieties of synthetic monomers and polymers are formed by the chemical modification of the oils.[71] The naturally occurring oils contain triglycerides chains having internal double bonds, alcohols, or epoxides which can be polymerized by different methods. The chemical modification of the triglycerides by incorporation of functional groups is one of the methods of polymerization. It is also observed from the literature that, the fatty acids have been used for the development of polymeric structures, both directly or as building blocks for the synthesis of complicated monomers.[72,73] Several varieties of materials such as linear, branched, and cross-linked polymers can be prepared from plant oils which can be used for the production of elastomers, rubbers, and composites. The polymer prepared from the triglyceride oils are polyesters, polyurethanes, polyamides, epoxy resins, polyesteramides etc. Among these plant oils, tung oil is readily available as a major product from the seeds of the tung tree, which is originally planted mostly in Asian countries, including China and some parts of Japan. This triglyceride of tung oil composed of 84% alpha-elaeostearic acid, 15% oleic acid or other unsaturated acids, and 5% saturated acids. The chemical structure of the tri-glyceride esters of fatty acids of natural oil is given in Figure 6.3. The oil is used as a drying oil in the preparation of paints and varnishes at room temperature due to its high unsaturation and conjugation of the chain.[74] The high degree of unsaturation of tung oil has prompted researchers to examine it as a potential monomer for free-radical polymerization or copolymerization into useful bulk polymers. It has been reported that the monocyclic dimeric fatty acid are produced from the dimerization of the elaeostearic acid by thermal polymerization of tung oil.[75] The synthesis of interpenetrating polymer networks (IPN) and a series of cross-linked copolymers are carried out from epoxidized vegetable oils and maleinized tung oil.[76] We have developed polymerized linseed oil-based sensor for the detection of volatile organic compounds such as o-xylene, p-cresol, benzene, and toluene.[77] Also, we have attempted to polymerize tung oil in the presence of benzoyl peroxide initiator, and the polymer film is coated on the quartz crystal microbalance surface for the detection of amine vapors.

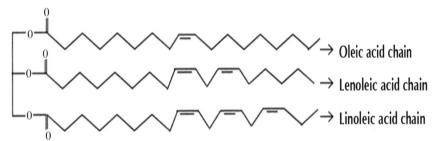

FIGURE 6.3 The chemical structure of the tri-glyceride esters of fatty acids of natural oil.

Quartz crystal microbalance (QCM) is a mass sensitive device. It is based on the principle of interactions between the coated surface of a quartz crystal and the analytes.[78] QCM is highly sensitive to mass changes on the nanogram scale and is extremely stable. It can work in the vacuum or liquid environment, and it is easy to measure small masses. The frequency shift is directly proportional to the adsorbed mass on the QCM. The correlation between the mass and frequency can be achieved by the Sauerbrey equation as stated above. The sensitivity of the sensor is determined by the chemical structure and physical properties of polymeric coating and also on the nature of interaction between polymer coating and analyte molecules.[79] Hence, we have chosen this piezoelectric material in our experiment for the coating of the polymerized tung oil for the detections of the organic vapors.

6.5.2 EXPERIMENTAL

6.5.2.1 MATERIALS AND SENSOR FABRICATION

Benzoyl peroxide, chloroform, methylamine, ethylamine, tert-butylamine, diethylamine, triethylamine, and ammonia are purchased from E. Merck, India. All the chemicals are used without any further purification. Tung oil is purchased from Sigma-Aldrich. The 10 MHz AT-cut quartz crystal microbalance crystals with silver electrodes on both sides are purchased from local market. A laboratory made set up is used for the frequency measurement of the quartz crystal. A set of six sensors is fabricated by varying the concentration of initiator in the solution. The concentrations of benzoyl peroxide used for this experiment are 0.34, 0.67, 1.01, 1.35, 1.69, and 2.08 (w/v %). Before coating of polymer film, the quartz crystal microbalances are rinsed with ethanol followed by deionized water. The sensors are fabricated with a maximum load of 5–6 kHz by simple solution dip–dry method. The solvent evaporation from the QCM surface is done at room temperature for 1 h, followed by 3-h drying at hot air oven at 110°C in the presence of argon. The optimization of the polymerization process is carried out and the plot is given in Figures 6.4 and 6.5. It is observed that the sensor fabricated with the solution containing 1.35 w/v% of benzoyl peroxide is more sensitive for the detection of volatile organic compounds.

Polymerization of Natural Oils

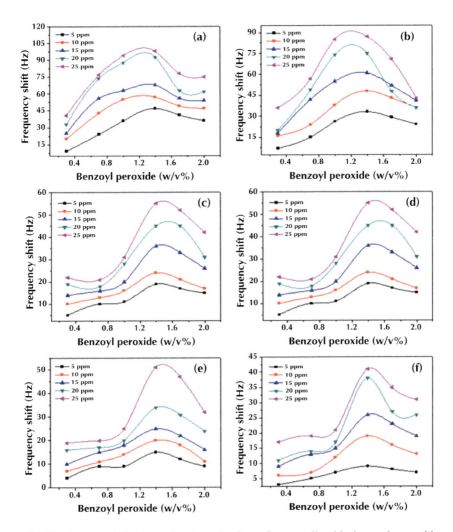

FIGURE 6.4 Optimization of polymerization of tung oil with benzoyl peroxide concentration in response to (a) methylamine, (b) ethylamine, (c) tert-butylamine, (d) diethylamine, (e) triethylamine, and (f) ammonia vapors (5–25 ppm).

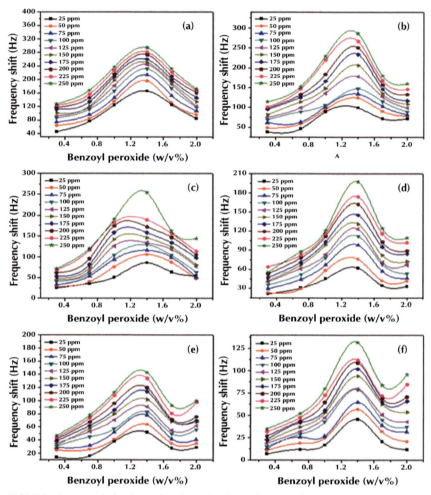

FIGURE 6.5 Optimization of polymerization of tung oil with benzoyl peroxide concentration in response to (a) methylamine, (b) ethylamine, (c) tert-butylamine, (d) diethylamine, (e) triethylamine, and (f) ammonia vapors (25–250 ppm).

6.5.2.2 EXPERIMENTAL SET UP

A typical quartz crystal microbalance system is consisted of quartz crystal, oscillator, frequency counter, and a computer. When nanogram of mass is deposited on the surface of the crystal, the thickness increases, consequently, the frequency of oscillation decreases from the initial value. In this experiment, the fabricated sensor is placed in a Teflon chamber of 100 mL

Polymerization of Natural Oils 149

volume at room temperature. The organic amines such as methylamine, ethylamine, tert-butylamine, diethylamine, triethylamine, and ammonia of different concentrations are injected into the sensor chamber using a glass syringe through a rubber septa attached to one end of the valve. A syringe pump (model no-LPM 50DN) is used to maintain the flow rate of the gas into the sensor chamber. The purging of sensor chamber is done by a continuous flow of dry nitrogen into the sensor chamber. A suction pump is attached to the sensor chamber to exhaust the analyte gases outside the chamber in small time duration. The purging is continued till the recovery of the baseline. The sensors are tested for in the concentration range of 5–250 ppm of volatile organic amines.

The structure of the polymerized tung oil film is characterized by Fourier transform infrared (FTIR) spectra (KBr dispersed pellets) in the range of 400–4000 cm^{-1} (model Paragon-500 FTIR of PerkinElmer spectrometer). The surface roughness of the QCM sensor surface before and after absorption of analyte is calculated from the atomic force microscope. A Pico plus 5500 ILM AFM (Agilent Technologies, USA) with a piezo scanner of maximum range of 100 μm is used for the measurement and Pico view software version 1.12 (Agilent Technologies, USA) is used for image processing.

6.5.3 RESULTS AND DISCUSSIONS

The FTIR spectrum of the polymerized film is given in Figure 6.6. The peaks at 1166 cm^{-1} and 1250 cm^{-1} are for –C–O stretching and –C–O stretching, respectively, and 1455 cm^{-1} and 2854 cm^{-1} are for –C–H bending and –C–H stretching, respectively. The peaks at 1645 cm^{-1} and 1745 cm^{-1} are for –C=C and –C=O stretching mode of vibration, respectively. Also the peaks at 2925 cm^{-1}, 3010 cm^{-1}, and 3550 cm^{-1} are for –C–H stretching, =C–H stretching, and –OH bending, respectively. The –OH group is incorporated to the triglycerides chain during the free radical polymerization of benzoyl peroxide. It is observed that the absorption of analyte is repeatable as the peaks in IR spectra do not change after repeated exposures of the film to amine vapors.

Figure 6.7 shows the AFM images before and after absorption of amine vapors. The AFM surface morphologies of the thin film before and after absorption of analytes are studied. It is observed that there is no appreciable change in the QCM surface after repeated experiments. The surface roughness slightly increases, which are due to the swelling of the polymer film. The sensor response to the amine vapors of concentration range 5–250 ppm are given in Figures 6.8 and 6.9. The sensor response is found to be linear to the concentration of benzoyl peroxide in solution. The frequency shift is more for methylamine vapor as compared to the other vapor due to its smaller molecular weight. The affinities toward amines are due to the formation of the hydrogen bonding and Vander walls force of attraction between the amine group of the analytes and the hydroxyl group in the cross-linked polymer film.

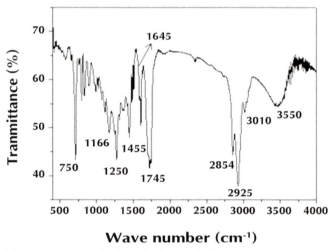

FIGURE 6.6 FTIR spectra of polymerized tung oil film.

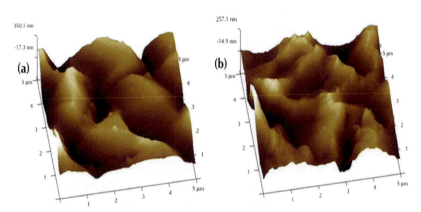

FIGURE 6.7 AFM images of the sensor (a) before and (b) after adsorption of analytes.

Polymerization of Natural Oils 151

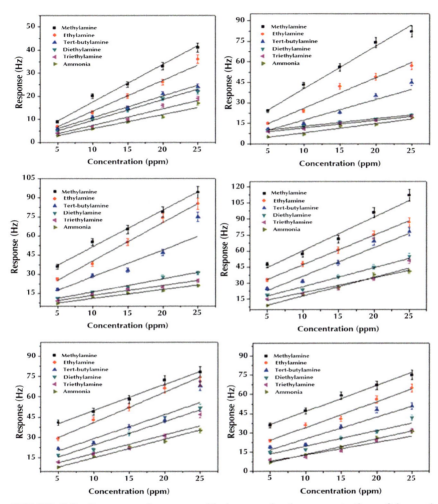

FIGURE 6.8 Response of sensors with increase in the concentration of benzoyl peroxide in solution (a) when exposed to 0.34, 0.67, 1.01, 1.35, 1.69 and 2.08% (w/v) to methylamine, ethylamine, tert-butylamine, diethylamine, triethylamine and ammonia vapors (5–25 ppm).

The sensor fabricated with the concentration of benzoyl peroxide higher than 1.35% (w/v) in solution shows decreases in sensitivity due to the formation of more rigid cross-linked polymer film on the QCM surface.

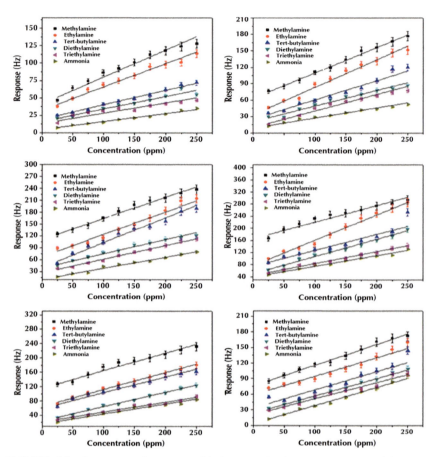

FIGURE 6.9 Response of sensors with increase in the concentration of benzoyl peroxide in solution (a) when exposed to 0.34, 0.67, 1.01, 1.35, 1.69 and 2.08% (w/v) to methylamine, ethylamine, tert-butylamine, diethylamine, triethylamine and ammonia vapors (25–250 ppm).

6.6 CASE STUDY 2: POLYMERIZED LINSEED OIL BASED QUARTZ CRYSTAL MICROBALANCE SENSOR AND ITS APPLICATION

6.6.1 INTRODUCTION

We have developed polymerized linseed oil-based quartz crystal microbalance (QCM) sensors for the detection of volatile organic compounds (VOCs) such as p-cresol, o-xylene, benzene, and toluene. The polymerization of

Polymerization of Natural Oils 153

linseed oil is carried out by using benzoyl peroxide as initiator followed by heating at 100°C. Simple solution dip–dry method is used for coating the polymer film on the crystal surface. The sensitivity, stability, and selectivity of the sensors are measured by the exposure of the QCM surface to vapors. The syntheses of polymers are optimized and it is found that the polymer prepared from 1.7% (w/v) of benzoyl peroxide gives good result in the detection of VOCs. The polymeric film is characterized by FTIR analysis and the surface morphologies of the film before and after absorption of vapor is analyzed by AFM. The sensor responses of concentration ranges from 5 to 250 ppm are found to be linear. The sensors are reactivated by releasing the adsorbed vapor by passing dry nitrogen gas. The VOCs like benzene, toluene, and xylene are produced in our daily lives from many sources like home and personal care products, building materials, and cosmetics. Generally, cresols are widely spread in nature and occur in many plants, petroleum, coal tar etc., and are emitted from municipal incinerators, combustion of coal and wood, vehicle exhaust, oil refineries etc.[80] Hence, the development of sensors for the detection of VOCs is needed for its potential applications in the fields of environmental and health monitoring.[81] Here, we are using quartz crystal microbalance (QCM) system for the detection of low concentration of organic volatiles such as benzene, toluene, o-xylene, and p-cresol. It has a unique resonance frequency and it converts surface acoustic waves to electric signals.

6.6.2 EXPERIMENTAL

Benzene, toluene, o-xylene, p-cresol, chloroform, and benzoyl peroxide are procured from E. Merck, India and linseed oil is procured from Sigma-Aldrich. All solvents and reagents are of analytical grade and used as it is without any further purification. Quartz crystals of frequency 10 MHz with silver electrodes on both sides are procured from local market. The laboratory-made instrument setup is used for frequency measurement.

6.6.3 RESULT AND DISCUSSION

The eight precursor solution is prepared by taking the benzoyl peroxide concentration 0.33, 0.68, 1.01, 1.35, 1.67, 2.00, 2.33, and 2.65% (w/v) in chloroform. The maximum loading of polymer film on QCM surface is approximately 5–6 KHz. The crystals are dried at room temperature followed by heating at 100°C for 1 h. The fabricated crystal is placed in the sensor chamber and the frequency of the crystal is measured. The purging of the sensor chamber is carried out by dry nitrogen. The VOCs of concentration range from 5–250 ppm is injected to the sensor chamber through glass syringe.

It is found that, sensor responses of the polymer-films for p-cresol, o-xylene, benzene, and toluene vapors are maximum when the polymer is formed with 1.67% (w/v) of benzoyl peroxide in solution. After that concentration, the sensor response decreases due to formation of more rigid structure (Fig. 6.10). The FTIR spectrum is shown in Figure 6.11.

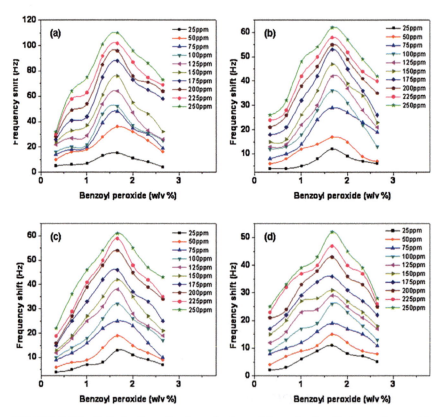

FIGURE 6.10 Optimization of polymerization of linseed oil with benzoyl peroxide concentration in response to (a) p-cresol, (b) o-xylene, (c) benzene, and (d) toluene vapors (25–250 ppm) (Ref. [77], Fig. 6.6).

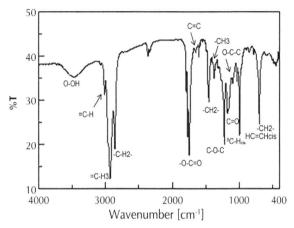

FIGURE 6.11 FT-IR spectra of polymerized linseed oil film. (Ref. [77], Fig. 6.3).

The spectra show the characteristic peaks at 1741 cm^{-1} (O–C–O), 1095 cm^{-1} (C–O), 606 cm^{-1} (HC–CHcis), 2855 cm^{-1}(–CH$_2$), 2928 cm^{-1} (–CH$_3$), 3447 cm^{-1} (O–OHstr), 1248 cm^{-1} (C–O–C), 1034 cm^{-1} (–C Hcis out of plane), and 1100 cm^{-1} (O–C–C) of the linseed oil. It is confirmed from the spectra that there is no appreciable change in the structure of linseed oil after polymerization. Sensor response in the concentration range from 25 to 250 ppm of p-cresol, o-xylene, benzene, and toluene is given in Figure 6.12. The maximum sensitivity is found with p-cresol and frequency shift of 110 Hz is obtained for 250 ppm of p-cresol vapor in presence of nitrogen atmosphere. VOCs of different concentrations are injected into the sensor chamber at 15-min interval. The repeatability of the sensor is tested and shown in Figure 6.13.

Thermal decomposition of benzoyl peroxide takes place to form free radicals, which helps for the propagation of the polymerization reaction. Auto-oxidation of linseed oil is also possible by using atmospheric air, which is a slow process. To speed up the reaction, previously some catalysts were used, for example, Mn or Co napthionates. It has been observed that, polymerized linseed oil-based QCM sensor has high frequency shift as compared to other vapors, it may be due to the polar interaction between the OH group of p-cresol with the ester groups of triglyceride present in the polymer chain. All the four types of VOCs at different concentrations, ranging from 25 to 250 ppm have similar characteristics. The sensitivity value of p-cresol vapor is found to be 0.41 Hz/ppm.

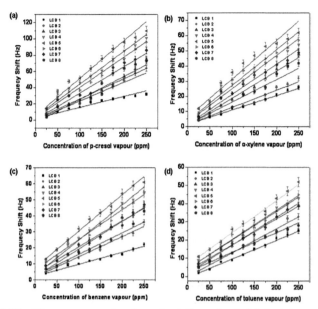

FIGURE 6.12 The response of polymerized linseed oil fabricated sensors to the concentration of (a) p-cresol, (b) o-xylene, (c) toluene, and (d) benzene vapors (Ref. [77], Fig. 6.8).

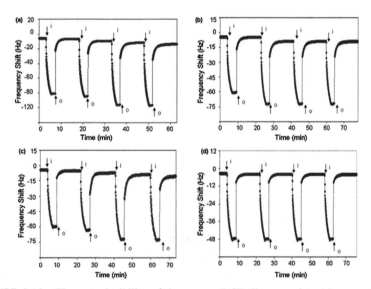

FIGURE 6.13 The reproducibility of the sensor (LCB 5) exposed to (a) p-cresol, (b) o-xylene, (c) toluene, and (d) benzene vapors of 250 ppm: VOC injection (i) and VOC desorption (o) by purging N_2 through the chamber (Ref. [77], Fig. 6.9).

6.7 CONCLUSION

In this study, we have developed polymerized tung oil- and polymerized linseed oil-based QCM sensor for the detection of volatile organic vapors. The polymeric film of tung oil shows higher sensitivity toward aliphatic amine vapor. The incorporation of –OH group in the triglyceride chain is confirmed from the FTIR spectra. The affinity toward amine vapor is due to the Van der Waals force of attraction and the hydrogen bonding between the –OH group and the amine group. From the AFM morphology study, it is concluded that the sensors are reproducible in nature. The sensors are activated by purging with dry nitrogen. The amines are present mostly in fish, semen, urine, etc. Also, aliphatic amines are industrially used for the production of dyes, corrosion inhibitor, and emulsifiers. It causes a health hazard directly and indirectly to the people working in these industries. Amine gases liberated from the fish or meat indicates its degree of spoilage. Hence, the detection of aliphatic amine vapor in ppm level is most important for environmental monitoring and food quality estimation. Similarly, the polymer film produced from linseed oil with benzoyl peroxide as the initiator is a useful polymer for the detection of volatile organic compounds. The sensitivity of the sensor increases linearly with benzoyl peroxide concentration up to 1.67% (w/v), which is due to the increase of cross-linked network in the polymer chain and the increase in free volume. After this concentration, the rigidity of the film increases, that is confirmed as the sensitivity decreases. The sensors are reactivated and reused after the release of the adsorbed vapor by passing dry N_2 gas into the sensor chamber. Hence, tung oil and linseed oil can be polymerized easily by using benzoyl peroxide and the polymer film is a useful film for gas sensor application.

KEYWORDS

- polymers
- natural oils
- polymerization
- free radical
- quartz crystal microbalance

REFERENCES

1. Meier, M. A. R.; Metzger, J. O.; Schubert, U. S. Plant Oil Renewable Resources as Green Alternatives in Polymer Science. *Chem. Soc. Rev.* 2007, *36*, 1788–1802.
2. Espinosa, L. M. de.; Michael, A.; Meier, R. Plant Oils: The Perfect Renewable Resource for Polymer Science. *Eur. Polym. J.* **2011**, *47*, 837–852.
3. Güner F. S., Yağcı, Y., Tuncer Erciyes, A. Polymers from Triglyceride Oils. *Prog. Polym. Sci.* **2006**, *31*, 633–670.
4. Erhan, S. Z.; Adhvaryu, A. Vegetable Based Base Stocks. In *Biobased Industrial Fluids and Lubricants;* AOCS Press; Cham-paign: IL, 2003; pp 1–19.
5. Meneghetti, S. M. P.; de Souza R. F.; Monteiro A. L.; de Souza M. O. Substitution of Lead Catalyst by Zirconium in the Oxidative Polymerization of Linseed Oil. *Prog. Org. Coat.* **1998**, *33*, 219–224.
6. Thames, S. F.; Wang, Z.; Brister, E. H.; Hariharan, R.; King, C. L.; Panjanani, K. G. Internally Plasticized and Low VOC Latex Compositions and Applications Thereof. U.S. Patent, 6,624,223, 2003.
7. Tortorello, A. J.; Montgomery, E.; Chawla, C. P. Radiation-Curable Compositions Comprising Oligomers Having an Alkyd Backbone. U.S. Patent, 6,638,616, 2003.
8. Motawie, A. M.; Hassan, F. A.; Manich, A.; Aboul-Fetouh, M. E.; El-din, A. Fakhr. Some Epoxidized Polyurethane and Polyester Resins Based on Linseed Oil. *J. Appl. Polym Sci.* **1995**, *55*, 1725–1732.
9. Gultekin, M.; Beker, U.; Guner, F. S.; Erciyes, A. T.; Yagci, Y. Styrenation of Castor Oil and Linseed Oil by Macromer Method. *Macromol. Mater. Eng.* **2000**, *283*, 15–20.
10. Akbas, T.; Beker, U. G.; Guner, F. S.; Erciyes, A. T.; Yagci, Y. Drying and Semidrying Oil Macromonomers III Styrenation of Sunflower and Linseed Oil. *J. Appl. Polym. Sci.* **2003**, *88*, 2373–2376.
11. Li, S. K. L.; Guillet, J. E. Photochemistry of Ketone Polymers: Photodegradation of an Amorphous Ethylene-Propylene Copolymer. *Macromolecules.* **1984**, *17*, 41–50.
12. Geuskens, G.; Kabamba, M. S. Photo-Oxidation of Polymers—Part V: A New Chain Scission Mechanism in Polyolefins. *Polym. Deg. Stab.* **1982**, *4*, 69–76.
13. Yin, Y.; Yao, S.; Zhou, X. Synthesis and Dynamic Mechanical Behavior of Cross-linked Copolymers and IPNs from Vegetable Oils. *J. Appl. Polym. Sci.* **2003**, *88*, 1840–1842.
14. Gultekin, M.; Beker, U.; Guner, F. S.; Erciyes, A. T.; Yagci, Y. Styrenation of Castor Oil and Linseed Oil by Macromer Method. *Macromol. Mater. Eng.* **2000**, *283*, 15–20.
15. Cassidy, P. E.; Schwank. Copolymerization of Dehydrated Castor Oil with Styrene: Determination of Reactivity Ratios. *J. Appl. Polym. Sci.* 1974, *18*, 2517–2526.
16. Devia, N.; Manson, J. A.; Sperling, L. H.; Conde, A. Simultaneous Interpenetrating Networksbased on Castor Oil Elastorners and Polystyrene. IV. Stress-strain and Impact Loading Behavior. *Polym. Eng. Sci.* 1979, *19*, 878–882.
17. Devia, N.; Manson, J. A.; Sperling, L. H.; Conde, A. Simultaneous Interpetrating Network Based on Castor Oil Elastomers and Polystyrene. III. Morphology and Glass Transition Behavior. *Polym. Eng. Sci.* 1979, *19*, 869–877.
18. Devia, N.; Manson, J. A.; Sperling, L. H.; Conde, A. Simultaneous Interpenetrating Networks Based on Castor Oil Elastomers and Polystyrene. Synthesis and Systems Characteristics. *Macromolecules.* 1979, *12*, 360–369.

Polymerization of Natural Oils 159

19. Kinabrew, R. G. Tung Oil in Mississippi. In *The Competitive Position of the Industry;* University of Mississippi: Oxford, MS, 1952.
20. Li, F.; Larock, R. C. Synthesis Structure and Properties of New Tung Oilstyrene-Divinylbenzene Copolymers Prepared by Thermal Polymerization. *Biomacromolecules.* 2003, *4,* 1018–1025.
21. Li, F.; Larock, R. C. Thermosetting Polymers from Cationic Copolymerization of Tung Oil: Synthesis and Characterization. *J. Appl. Polym. Sci.* **2000,** *78,* 1044–1056.
22. Trumbo, D. L.; Mote, B. E. Synthesis of Tung Oil–Diacrylate Copolymers via the Diels-Alder Reaction and Properties of Films from The Copolymers. *J. Appl. Polym. Sci.* **2001,** *80,* 2369–2375.
23. Chen, X.; Wudl, F.; Mal, A. K.; Shen, H.; Nutt, S. R. New Thermally Remendable Highly Cross-Linked Polymeric Materials. *Macromolecules.* 2003, *36,* 1802–1807.
24. Mercango, Z. M.; Kusefog, lu. S.; Akman, U.; Hortac, su. O. Polymerization of Soybean Oil *via* Permanganate Oxidation with Sub/Supercritical CO_2. *Chem. Eng. Process.* **2004,** *43,* 1015–1027.
25. Braunegg, G.; Lefebvre, G.; Genser, K. F. Polyhydroxyalkanoates, Biopolyesters from Renewable Resources: Physiological and Engineering Aspects. *J. Biotechnol.* **1998,** *65,* 127–161.
26. Ashby, R. D.; Cromwick, A.; Foglia, T. A. Radiation Crosslinking of a Bacterial Medium-Chain-Length Poly(hydroxyalkanoate) Elastomer from Tallow. *Int. J. Biol. Macromol.* 1998, *23,* 61–72.
27. Poirier, Y. Production of New Polymeric Compounds in Plants. *Curr. Opin. Microbiol.* 1999, *10,* 181–185.
28. Van der Walle, G. A. M.; Buisman, G. J. H.; Weusthuis, R. A.; Eggink, G. Development of Environmentally Friendly Coatings and Paints Using Medium-Chain-Length Poly(3-hydroxy lkanoates) as the Polymer Binder. *Int. J. Biol. Macromol.* **1999,** *25,* 123–128.
29. Ashby, R. D.; Foglia, T. A.; Solaiman, D. K. Y.; Liu, C.; Nunez, A.; Eggink, G. *Int. J. Biol. Macromol.* **2000,** *27,* 355–361.
30. Eggink, G.; de Waard, P.; Huijberts, G. N. M. Can. Formation of Novel Poly (hydroxyalkanoates) from Long-Chain Fatty Acids. *J. Microbial.* **1995,** *41,* 14–21.
31. Ashby, R. D.; Cromwick, A.; Foglia, T. A. Radiation Crosslinking of a Bacterial Medium-Chain-Length Poly(hydroxyalkanoate) Elastomer from Tallow. *Int. J. Biol. Macromol.* 1998, *23,* 61–72.
32. Cromwick, A. M.; Foglia, T.; Lenz, R. W. The Microbial Production of Poly (hydroxyalkanoates) from Tallow. *Appl. Microbiol. Biotechnol.* **1996,** *46,* 464–469.
33. Ashby, R. D.; Foglia, T. A. Poly(hydroxyalkanoate) Biosynthesis from Triglyceride Substrates. *Appl. Microbial. Biotechnol.* 1998, *49,* 431–437.
34. Hazer, B.; Demirel, S. I.; Borcakli, M.; Eroglu, M. S.; Cakmak, M.; Erman, B. Free Radical Crosslinking of Unsaturated Bacterial Polyesters Obtained from Soybean Oily Acids. *Polym. Bull.* **2001,** *46,* 389–394.
35. Casini, E.; deRijk, T. C.; deWaard, P.; Eggink, G. Synthesis of Poly(hydroxyalkanoate) from Hydrolyzed Linseed Oil. *J. Environ. Polym. Degrad.* 1997, *5,* 153–158.
36. Yeganeh, H.; Mehdizadeh, M. R. Synthesis and Properties of Isocyanate Curable Millable Polyurethane Elastomers Based on Castor Oil as a Renewable Resource Polyol. *Eur. Polym. J.* 2004, *40,* 1233–1238.

160 Functional Polymeric Composites: Macro to Nanoscales

37. Kaplan, D. L. *Biopolymers from Renewable Resources;* Springer: New York, 1998.
38. Li, F.; Larock, R. C. Synthesis, Structure and Properties of New Tung Oil−Styrene−Divinylbenzene Copolymers Prepared by Thermal Polymerization. *Biomacromolecules.* **2003,** *4,* 1018–1025.
39. C- akmakli, B.; Hazer, B.; Tekin, I. O. Beg endik Co, mert. Synthesis and Characterization of Polymeric Soybean Oil-G-Methyl Methacrylate (and n-butyl methacrylate) Graft Copolymers: Biocompatibility and Bacterial Adhesion. *Biomacromolecules.* **2005,** *6,* 1750–1758.
40. C- akmakli, B.; Hazer, B.; Tekin, I. O.; Kizgut, S.; Koksal, M.; Menceloglu, Y. Synthesis and Characterization of Polymeric Linseed Oil-G-Methyl Methacrylate or Styrene. *Macromol. Biosci.* **2004,** *4,* 649–665.
41. Crivello, J. V.; Narayan, R. Epoxidized Triglycerides as Renewable Monomers in Photoinitiated Cationic Polymerization. *Chem. Mater.* 1992, *4,* 692–699.
42. Chakrapani, S.; Crivello, J. V. Synthesis and Photoinitiated Cationic Polymerization of Epoxidized Castor Oil and Its Derivatives. *J. Macromol. Sci.—Pure Appl. Chem.* 1998, *A35,* 251–262.
43. Crivello, J. V.; Carlson, K. D. Photoinitiated Cationic Poly- Merization of Naturally Occurring Epoxidized Triglycerides. *J. Macromol. Sci. Pure Appl. Chem.* **1996,** *A33,* 691–710.
44. Thames, S. F.; Yu, H. Cationic UV-Cured Coatings of Epoxidecontaining Vegetable Oils. *Surf. Coat. Technol.* **1999,** *115,* 208–214.
45. Gu, H. Y.; Ren, K.; Martin, D.; Marino, T.; Neckers, D. C. Cationic UV-Cured Coatings Containing Epoxidized Soybean Oil Initiated by New Onium Salts Containing Tetrakis(pentafluorophenyl)Gallate Anion. *J. Coat. Technol.* **2002,** *74,* 49–52.
46. Wan Rosli, W. D.; Kumar, R. N.; Mek Zah, S.; Mohd, H. M. UV Radiation Curing of Epoxidized Palm Oil-Cycloaliphatic Diepoxide System Induced by Cationic Photoinitiators for Surface Coatings. *Eur. Polym. J.* **2003,** *39,* 593–600.
47. Zou, K.; Soucek, M. D. UV-Curable Cycloaliphatic Epoxide Based on Modified Linseed Oil: Synthesis, Characterization and Kinetics. *Macromol. Chem. Phys.* **2005,** *206,* 967–975.
48. Park, S., J.; Jin, F. L.; Lee, J. R. Synthesis and Thermal Properties of Epoxidized Vegetable Oil. *Macromol. Rapid Commun.* **2004,** *25,* 724–727.
49. Park, S. J.; Jin, F. L.; Lee, J.; Shin, J. S. Cationic Polymerization and Physicochemical Properties of a Biobased Epoxy Resin Initiated by Thermally Latent Catalysts. *Eur. Polym. J.* **2005,** *41,* 231–237.
50. Potter, T. A.; Williams, J. L. Coatings Based on Polyurethane Chemistry—an Overview and Recent Developments. *J. Coat. Technol.* **1987,** *59,* 63–72.
51. Alam, M.; Sharmin, E.; Ashraf, S. M.; Ahmad, S. Newly Developed Urethane Modified Polyesteramide-Based Anticorrosive Coatings from a Sustainable Resource. *Prog. Org. Coat.* **2004,** *50,* 224–230.
52. Guang, L.; Gaymans, R. J. Polyesteramides with Mixtures of Poly(tetramethylene oxide) and 1,5-Pentandiol. *Polymer.* **1997,** *38,* 4891–4896.
53. Mendes, R. K.; Claro-Neto, S.; Cavalheiro, E. T. G. Evaluation of a New Rigid Carbon-Castor Oil Polyurethane Composite as an Electrode Material. *Talanta.* **2002,** *57,* 909–917.

54. Dweib, M. A.; Hu, O'Donnell, A.; Shenton, H. W.; Wool, R. P. All Natural Composite Sandwich Beams for Structural Applications. *Compos. Struct.* **2004**, *63*, 147–157.
55. O'Donnell, A.; Dweib, M. A.; Wool, R. P., Natural Fiber Composites with Plant Oil-Based Resin. *Compos. Sci. Technol.* **2004**, *64*, 1135–1145.
56. O'Sillivan, C. K.; Guibault, G. G. Commercial Quartz Crystal Microbalances-Theory and Applications. *Biosens. Bioelectron.* **1999**, *14*, 663–670.
57. Sakai, G.; Sakai, T.; Uda, T.; Miura, N.; Yamazoe, N. Evaluation of Binding of HAS to Monoclonal and Polyclonal Antibody by PZ Immunosensing. *Sens. Actuators B.* **1995**, *42*, 89–94.
58. Carter, R. M.; Jacobs, M. B.; Lubrano, G.; Guilbault, G. G. Piezoelectric Detection of Ricin and Affinity Purified Goat Anti-Ricin. *Anal. Lett.* **1995**, *28*, 1379–1386.
59. Carter, R. M.; Mekalanos, J. J.; Jacobs, M. B.; Lubrano, G. J.; Guilbault, G. G. Quartz Crystal Microbalance Detection of Vibriocholerase. *J. Immunol. Methods.* **1996**, *187*, 121–125.
60. Harteveld, J. L. N.; Nieuwenhuizen, M. S.; Wils, E. R. J. Detection of Staphylococcal Enterotoxin B Employing a Pzimmunosensor. *Biosens. Bioelectron.* **1997**, *12*, 661–667.
61. Li, B.; Sauve, G.; Iovu, M. C.; Jeffries-El, M.; Zhang, R.; Cooper, J.; Santhanam, S.; Schultz, L.; Revelli, J. C.; Kusne, A. G., Kowalewski, T.; Snyder, J. L.; Weiss, L. E.; Fedder, G. K.; McCullough, R. D.; Lambeth, D. N. Volatile Organic Compound Detection Using Nanostructured Copolymers. *Nano Lett.* **2006**, *6*, 1598–1602.
62. Dua, V.; Surwade, S. P.; Ammu, S.; Agnihotra, S. R.; Jain, S.; Roberts, K. E.; Park, S.; Ruoff, R. S.; Manohar, S. K. All-Organic Vapor Sensor Using Inkjet-Printed Reduced Graphene Oxide. *Angew. Chem. Int. Edn.* **2010**, *49*, 2154–2157.
63. Schlupp, M.; Weil, T.; Berresheim, A. J.; Wiesler, U. M.; Bargon, J.; Mullen, K. Polyphenylene Dendrimers as Sensitive and Selective Sensor Layers. *Angew. Chem. Int. Edn.* **2001**, *40*, 4011–4015.
64. Chen, X.; Parker, S. G.; Zou, G.; Su, W.; Zhang, Q. J. Beta-Cyclodextrin-Functionalized Silver Nanoparticles for the Naked Eye Detection of Aromatic Isomers. *ACS Nano.* **2010**, *4*, 6387–6394.
65. Tancrede, M.; Wilson, R.; Zeise, L.; Crouch, E. A. C. The Carcinogenic Risk of Some Organic Vapors Indoors: A Theoretical Survey. *Atmos. Environ.* **1987**, *21*, 2187–2205.
66. Molhave, L. Indoor Climate, Air Pollution, and Human Comfort, *J. Expo. Anal. Environ. Epidemiol.* **1991**, *1*, 63–81.
67. Huang, C. Y.; Song, M.; Gu, Z. Y.; Wang, H. F.; Yan, X. P. Probing the Adsorption Characteristic of Metal–Organic Framework MIL-101 for Volatile Organic Compounds by Quartz Crystal Microbalance. *Environ. Sci. Tech.* **2011**, *45*, 4490–4496.
68. Buszewski, B. A.; Kesy, M.; Ligor, T.; Amann, A. Human Exhaled Air Analytics: Biomarkers of Diseases. *Biomed. Chromato.* **2007**, *21*, 553–566.
69. Meier, A. R. M.; Metzger, J. O.; Schubert, U. S. Plant Oil Renewable Resources as Green Alternatives in Polymer Science. *Chem. Soc. Rev.* **2007**, *36*, 1788–1802.
70. Metzger, J. O. Fats and Oils as Renewable Feedstock for Chemistry. *Eur. J. Lipid Sci. Technol.* **2009**, *111*, 865–876.
71. Guner, F. S.; Yagcı, Y.; Erciyes, A. T. Polymers from Triglyceride Oils. *Polym. Sci.* **2006**, *31*, 633–670.

72. Sharma, V.; Kundu, P. Addition Polymers from Natural Oils—A Review. *Prog. Polym. Sci.* **2006,** *31,* 983–1008.
73. Sharma, V.; Kundu, P. A Review on Polymer–Layered Silicate Nanocomposites, *Prog. Polym. Sci.* **2008,** *33,* 1199–1215.
74. Kinabrew, R. G. Tung Oil in Mississippi. In *The Competitive Position of the Industry;* University of Mississippi: Oxford, MS, 1952.
75. Rheineck, A. E.; Austin, A. O. Myers Raymond R, Long J. S. *Drying Oils—Modification and Use: Treatise on Coatings Part II;* Marcel Dekker: New York, 1967; Vol. 1.
76. Yin, Y.; Yao, S.; Zhou, X., Synthesis and Dynamic Mechanical Behavior of Crosslinked Copolymers and IPNs from Vegetable Oils, *J. Appl. Polym. Sci.,* **2003,** *88,* 1840–1842.
77. Das, R.; Biswas, S.; Bandyopadhyay, R.; Pramanik, P. Polymerized Linseed Oil Coated Quartz Crystal Microbalance for the Detection of Volatile Organic Vapors. *Sens. Actuators B: Chem.* **2013,** *185,* 293–300.
78. Zhou, X. C.; Zhong, L.; Li, S. F. Y.; Ng, S. C.; Chan, H. S. O. Organic Vapor Sensors Based on Quartz Crystal Microbalance Coated with Self-Assembled Monolayers. *Sens. Actuators B: Chem.* **1997,** *42,* 59–65.
79. Mirmohseni, A.; Oladegaragoze, A. Determination of Chlorinated Aliphatic Hydrocarbons in Air Using a Polymer Coated Quartz Crystal Microbalance Sensor. *Sens. Actuators B: Chem.* **2004,** *102,* 261–270.
80. Amann, A.; Poupart, G.; Telser, S.; Ledochowski, M.; Schmid, A.; Mechtcheriakov, S. Applications of Breath Gas Analysis in Medicine. *Int. J. Mass Spect.* **2004,** *239,* 227–233.
81. Srivastava, A. K. Detection of Volatile Organic Compounds (VOCs) Using SnO_2 Gas-Sensor Array and Artificial Neural Network. *Sens. Actuators B: Chem.* **2003,** *96,* 24–37.

CHAPTER 7

EMERGENCE OF A NEW NANOMATERIAL: NANOCELLULOSE AND ITS NANOCOMPOSITES

CHI HOONG CHAN, CHIN HUA CHIA*, and SARANI ZAKARIA

Faculty of Science and Technology, Bioresources and Biorefinery Laboratory, School of Applied Physics, Universiti Kebangsaan Malaysia, Bangi 43600, Selangor, Malaysia

*Corresponding author. E-mail: chiachinhua@yahoo.com; chia@ ukm.edu.my

CONTENTS

Abstract 164

7.1 Introduction 164

7.2 Nanocellulose 165

7.3 Nanocomposites 178

7.4 Future Perspectives 186

7.5 Acknowledgments 186

Keywords 187

References 187

ABSTRACT

Biobased nanofibers, nanocelluloses, have been studied extensively and received a lot of attention recently due to its renewability, biodegradability, and superior mechanical properties. The production of nanocelluloses typically begins with physical, chemical, and enzymatic pre-treatment of biomass to remove lignin and hemicellulose from biomass sources to mitigate the high cost of nanocellulose production. There are three different types of nanocelluloses that can be extracted or produced, namely, cellulose nanocrystals (CNC), cellulose nanofibrils (CNF), and bacterial cellulose (BC). Each nanocellulose differs in its mechanical, physical, chemical properties, biomass sources, and its method of productions (pre-treatment and defibrillation method) are reviewed in this chapter. The multiple applications of nanocellulose with the focus on nanocomposites are also reviewed. The commercial production of nanocelluloses is expected to increase remarkably in near future.

7.1 INTRODUCTION

Nanocellulose is an emerging new class material extracted from cellulose. Its lateral dimension is less than <100 nm. There are three types of nanocelluloses, that is, cellulose nanofibrils (CNF), cellulose nanocrystals (CNC), and bacterial cellulose (BC).[1] These nanocelluloses differ in method of production, where, CNF is produced through delamination of cellulose, CNC through chemical extraction method and BC is synthesized from bacteria (*Acetobacter xylinum*).[1] Nanocellulose can be extracted from any cellulosic plant source after the removal of lignin and hemicelluloses (optional), followed by chemical extraction or mechanical disintegration.

Individual cellulose microfibril consists of crystalline and amorphous regions.[2] The crystalline region of cellulose, CNC, can be extracted and was found to be high in modulus.[3] Additionally, individual cellulose microfibril can also be extracted from cellulose pulp through mechanical means to obtain CNF.

Stable CNC was first extracted from cellulose using concentrated acid hydrolysis by Mukherjee et al. in 1953,[4] since then, various applications of using this versatile material have been proposed by researchers. Extracted CNC is of particular interest due to its unique properties, for example, CNC self-assemble to a chiral nematic phase at a certain onset concentration as observed by Dong et al.,[5] where, evaporated CNC suspension produces a solid iridescent film which retains its self-assembled chiral nematic state after drying.[6] The perceived color of iridescent CNC film depends on the pitch of cholesteric order and the angle of incidence of light.[7] The tunability of reflected polarized light relies on the processing which determines the arrangement of CNC which affects the photonic crystals behavior, that is, if the pitch of helix is between 400 and 700 nm, a visible iridescent color can be observed.[8] The arrangement are influenced by the aspect ratio of CNC,[9] sulfate ester content,[10] electrolytes content,[11] types of solvent used,[12] etc. However, the mechanisms of arrangement of chiral nematic phase during drying and explanation of iridescence effect are still under investigation and debate.

For CNF extraction, the process typically begins using readily available bleached kraft pulp or sulfite pulp.[13] The pulp can then undergo one of the various methods to defibrillate cellulose, namely, high-pressure homogenization,[14] steam explosion,[15] high speed blender,[16] etc. CNF was first extracted by Turbak et al. in 1983 by defibrillating wood pulp at high pressure using a Gaulin homogenizer.[17] Turbak et al. called the resulting suspension, microfibrillated cellulose (MFC).[17] However, the terms, CNF, MFC, nanofibrillated cellulose (NFC), cellulose microfibrils (MCF), and sometimes nanofibers are used interchangeably to refer to defibrillated wood pulp by physical means with or without chemical, enzymatic, or physical pre-treatment.[1,18] Defibrillation of cellulose is a very energy extensive process,[19] but the energy cost can be mitigated by pre-treatment. Pre-treatment usually aims to limit the hydrogen bonding, adding a repulsive charge on the surface of CNF and decreasing the degree of polymerization (DP).[20] A suitable pre-treatment is carboxylation, which involves oxidation of macrofibrils which induces carboxylation on the cellulose surface in order to facilitate the second step, disintegration. Carboxylation can be performed using 2,2,6,6-tetramethylpiperidine-1-oxyl radical (TEMPO) mediated oxidation,[21] enzymatic,[22] and acidified chlorite.[23] CNF has a wide and potential applications, such as, reinforcement in composites,[24] thickening agent in food and paint,[25] electronic devices,[26] and scaffold for tissue growth.[27]

7.2 NANOCELLULOSE

Nanocelluloses are cellulose in nano-size, that is, at least one dimension of the materials produced is in the range of nano-scale (<100 nm). According to Beck-Candanedo et al., in 1949, Rånby and Ribi were the first to publish their findings on hydrolyzing cellulose to produce a stable suspension of colloidal-sized cellulose, that is, CNC.[28] The production of CNF was first reported by Turbak et al. in 1983 by defibrillating cellulose using a Gaulin homogenizer.[17] Since then, various types of nanocelluloses and different methods of production had been employed. Numerous names are given to nanocellulose which are dependent on the method of production and the shapes of nanocelluloses obtained. However, sometimes, the names given have no particular order or rules as summarized and shown in Table 7.1. Nanocellulose can be grouped into three categories depending on its method of production and physical properties. Table 7.2 shows the characteristics of the three nanocelluloses and their naming. To avoid confusion, the terms CNC and CNF will be used throughout the chapter. Figure 7.1 shows electron micrographs of CNC and CNF produced from kenaf core wood.

TABLE 7.1 Various Names of Nanocellulose from Various Methods and Sources.

Name	Source	Method/Process	Source
Cellulose nano whiskers	Microcrystalline cellulose	H_2SO_4 hydrolysis	[29]
	Ramie	H_2SO_4 hydrolysis	[7]
	Microcrystalline cellulose	LiCl/DMAc	[30]
Cellulose nanocrystals	Whatman filter paper	H_2SO_4 hydrolysis	[31]
	Bacterial cellulose	H_2SO_4 hydrolysis	[32]
	Cotton	H_2SO_4 hydrolysis	[33]
Whiskers	Cellulose fibers	H_2SO_4 hydrolysis	[34]
Nanofibers	Wheat straw	HCl hydrolysis and mechanical disintegration	[35]
Microfibrillated cellulose	Pulp	Gaulin homogenization	[36]
Cellulose microfibrils	Pulp	Gaulin homogenization	[37]
Nanofibrillated cellulose	Sulfite pulp	Mechanical homogenization	[38]
Nanocellulose	Sisal	H_2SO_4 hydrolysis	[39]

Source: Adapted from Siqueira et al.[18]

Emergence of a New Nanomaterial

TABLE 7.2 Types of Nanocellulose.

Types of Nanocellulose	Cellulose Nanocrystals (CNC)	Cellulose Nanofibrils (CNF)	Bacterial Cellulose (BC)
Source	Wood, cotton, hemp, flax, wheat straw, mulberry bark, algae, and bacterial cellulose	Wood, beets, bamboo, hemp, and flax	Low molecular weight sugar and alcohol
Synonym Terms/name	Cellulose nanocrystals, rod-like crystalline cellulose, crystallite, crystalline whiskers, and microcrystals	Cellulose microfibril, defibrillated microcellulose, defibrillated nanocellulose, microfibril, and nanofibril	Microbial cellulose and bio-cellulose
Average dimensions	Diameter: 5–70 nm Length:100–250 nm (cellulose obtained from plants), >100 nm, up to few microns (cellulose obtained from tunicate, algae, and bacteria)	Diameter: 5–60 nm Length : up to few micrometers	Diameter: 20–100 nm
Method of production	Acidic hydrolysis from cellulose obtained from various sources	Delamination of wood pulp through mechanical with or without, enzymatic or chemical treatment	Bacterial synthesis

Source: Adapted from Klemm et al.[1]

Usually, cellulose is extracted from plant source before chemically hydrolyzed or physically delaminated to nano-size. First, cellulose is extracted using alkaline treatment or pulping to remove hemicelluloses and separate cellulose from the fibers. Bleaching or delignification is performed subsequently to remove residue lignin to obtain pure cellulose. The detailed extraction methods are discussed further below.

7.2.1 CELLULOSE NANOCRYSTALS (CNC)

Cellulose must be extracted from plant biomass prior to CNC extraction. Alkaline treatment is performed mainly to remove hemicelluloses and to swell the fibers, while bleaching or delignification is performed to remove lignin present in plant biomass.

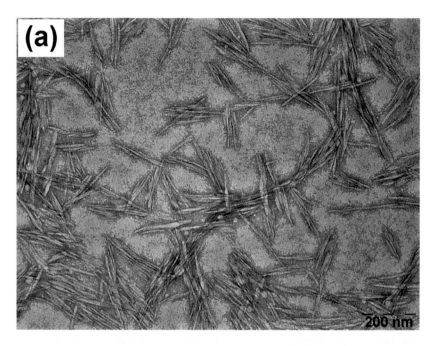

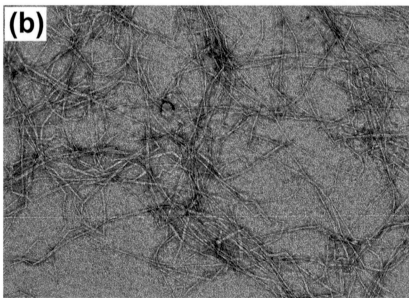

FIGURE 7.1 TEM image of CNC and CNF obtained from kenaf core wood: (a) CNC and (b) CNF.

Emergence of a New Nanomaterial

a) Alkaline Treatment

Besides removing hemicelluloses, alkali treatment removes fatty acids, alkaline soluble lignin, and other foreign matters (e.g., silicon dioxide). Alkaline treatment is usually performed by using hot sodium hydroxide (NaOH). An increasing alkali concentration and temperature has been shown to remove more hemicelluloses and silica.[40] Alkali (NaOH) also has a saponification function, where the breaking of alkali labile bonds, that is, ester linkages in lignin-carbohydrate allows the liberation of lignin.[41] The removal of hemicelluloses and alkali soluble lignin allows the separation of cellulose from the raw material. However, if the lignin content is still too high, it will impede further chemical treatment (acidic hydrolysis).[42]

b) Delignification

In pulp and paper industry, colored compounds in unbleached pulps mainly come from lignin which is undesirable. Hence, in order to bleach pulps, delignification can be performed and vice versa. Many parallels can be drawn between bleaching in pulps and delignification; however, it should never be assumed to be the same. The objective of bleaching is to remove color so as to whiten the pulp while delignification is to remove lignin. A whitened pulp does not necessarily remove all lignin, for example, due to sometimes inaccessible lignin in the pulp fiber. For the sake of discussion, bleaching in wood pulp will also be discussed.

For wood pulp, bleaching is performed by degrading any residual lignin. Due to the bond structures of lignin which are high in conjugated double bonds,[43] the brightness of pulp is associated with degree of degradation (oxidation) of lignin. Lachenal and Muguet categorized bleaching chemicals into three groups according to their reactivity toward residual lignin structures as shown in Table 7.3.[43,44] Lignin represents the hydrophobic part in the fibers which is undesirable in wood pulp and nanocellulose extraction. In CNC extraction for example, high lignin content is undesirable due to the inhibition of acidic hydrolysis caused by lignin.[42] A total removal of lignin in natural fibers is very difficult if not impossible.

As summarized above, an electrophilic bleachants, such as Cl_2, ClO_2, O_3, and O_2 will quickly degrade lignin by electrophilic addition of oxidant to the aromatic nuclei.[43] The most popular bleachant, ClO_2, shows a high selectivity toward oxidation of lignin. Oxidizing lignin will turn it

170 Functional Polymeric Composites: Macro to Nanoscales

water-soluble chlorinated lignin while further oxidation generally yield either quinones or muconic acids which are generally resistant to further oxidation.[45]

TABLE 7.3 Classification of Bleachants.

	Category		
	I	**II**	**III**
Bleachants			
Chlorine based	Cl_2	ClO_2	$NaOCl$
Chloride free	O_3	O_2	H_2O_2
Type of reaction	Electrophilic	Electrophilic	Nucleophilic
pH	acid	acid/alkaline	alkaline
Reaction sites in lignin structures	Olefinic and aromatic	Free phenolic groups and double bonds	Carbonyl groups and conjugated double bonds

Source: Adapted from Sixta et al.[43]

In CNC or CNF extraction particularly in a laboratory setting, sodium chlorite ($NaClO_2$) is used for delignification. ClO_2 can be produced by $NaClO_2$ degradation in hot acid solution. A popular method is Wise method, originally used to determine holocellulose content where natural fibers are delignified consecutively by adding $NaClO_2$ and acetic acid.[46] More recently, $NaClO_2$ is used but in an acetate buffer solution.[18]

c) Acidic Hydrolysis

The most common method for CNC extraction is via acidic hydrolysis. Prepared CNC is subjected to the type of acid used, acid concentration, temperature, and duration of hydrolysis. The most commonly used acid in CNC extraction is concentrated H_2SO_4 (64 wt%).[18] Other acids also had been used to extract CNC, namely, hydrochloric acid (HCl),[47] hydrobromic acid (HBr),[48] phosphoric acid (H_3PO_4),[49] acetic acid,[49] etc. CNC extracted from H_2SO_4 is a highly stable colloidal aqueous suspension due to esterified charged sulfate groups on its surface which gives a dual-layer electric repulsion effect.[50] On the other hand, using HCl produces a less stable CNC with minimal surface charge.[51]

CNC can be extracted from lignocellulosic material, for example, cotton linter, sisal, bamboo, kenaf bast,[52] kenaf core,[53] etc. Some CNC can also be extracted from exotic materials, such as *Valonia*[54] and tunicate, *Microcosmus fulcatus*.[55] Acidic hydrolysis relies on the degradation of amorphous region in cellulose, due to preferential hydrolysis of a more accessible material (amorphous cellulose) as shown in Figure 7.2. Model compound, microcrystalline cellulose were acid hydrolyzed with different hydrolysis time and were found that its length and surface charge became relatively stable after 1 h at 45°C with 64 wt% sulfuric acid (H_2SO_4).[5] Another study also shows that the crystallinity index (CrI) of CNC from kenaf bast cellulose peaked after 40 min of hydrolysis at 45°C with 64 wt% sulfuric acid.[52] This supports the model proposed by Millet et al., in which, easily accessible amorphous β (1 → 4) bonds in cellulose are hydrolyzed, what was left is the residual crystalline region in cellulose.[56]

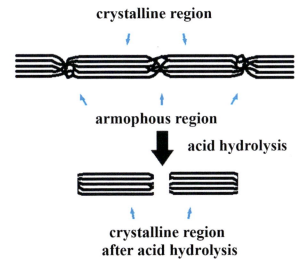

FIGURE 7.2 Acidic hydrolysis process shows the preferential hydrolysis of amorphous cellulose.

d) Mechanical and Physical Properties

CNC has superior mechanical properties (three times the specific modulus of structural steel),[57] non-toxic,[58] renewable, and biodegradable. Table 7.4 shows the moduli of some engineering materials, although the Young's modulus of CNC is lower than steel but CNC is five times lighter than steel. Although CNC is physically strong, due to its rich hydroxyl groups on its surface means it is highly hydrophilic.[59] On the other hand, H_2SO_4 extracted CNC is esterified with sulfate groups on its surface which made it negatively charged. This is unfavorable for reinforcing purposes as most polymer is hydrophobic hence, to improve compatibility, chemical modification, surfactant coating, or polymer grafting is required.[7]

TABLE 7.4 Moduli of Some Engineering Materials Compare to CNC.

Material	Young's Modulus (GPa)	Density (Mg m–3)	Specific Modulus (GPa Mg–1 m3)
Aluminum	69	2.7	26
Steel	200	7.8	26
Glass	69	2.5	28
CNC	138, 150.7 [3]	1.5	9

Source: Adapted from Eichhorn et al.[57]

Typically, the length of the CNC ranges from 100 to 250 nm depending on the sources and other factors stated earlier. Usually, transmission electron microscope (TEM) is employed to image extracted CNC while its length and lateral dimension are measured. CNC is assumed to be cylindrical in shape (elongated tabular) based on the earliest observation using electron microscope.[4] Cross sectional of *Valonia* CNC was found to be square in shape, thus the CNC is rectangular in shape.[54] While more recently, it was found to be ribbon in shape, that is, its lateral dimension differs from its thickness.[9] However, an ellipsoidal model of CNC proposed based on atomic force microscope (AFM) and TEM observations appears to fit well.[60]

7.2.2 CELLULOSE NANOFIBRILS (CNF)

CNF is produced by defibrillating extracted cellulose to fibrils. One of the earliest CNF produced is Turbak et al. in 1983, with diameter between 20 nm and 60 nm while length in a few microns.[17] Turbak et al. refer to their cellulose as MFC after undergoing defibrillation by Gaulin homogenizer.

Elementary fibril is the basic structure of cellulose in plant fiber, they are the universal structural units in natural cellulose. Elementary fibril in various plants such as ramie, cotton, *Valonia*, dates and ivory nuts were observed to be 3.5 nm in diameter.[61] Therefore, theoretically, a completely defibrillated cellulose will be 3.5 nm in lateral dimension. However, this is not the case where most of the CNF produced contained 10–50 elementary microfibrils.[20,62] On the other hand, chemical modified cellulose using TEMPO-oxidation method by Isogai et al. were found to produce a homogenous CNF with lateral dimension between 3 nm and 4 nm[13] which is very close to the lateral dimension of elementary fibrils. Recently, TEMPO-oxidized CNF produced from rice straw were found to be 1–2 nm as observed under TEM.[63] While TEMPO-oxidized CNF with extensive ultrasonic treatment yield CNF with height of <1 nm as observed by AFM.[64]

Therefore, in order to produce CNF, mechanical and physical disintegration is required to defibrillate (delaminate) cellulose. A two-step defibrillation is employed, that is, chemical or enzymatic pre-treatment follow by physical disintegration. Pretreatment and physical disintegration of cellulose to produce CNF are discussed below.

a) Pretreatment

The process typically begins using readily available bleached Kraft pulp or sulfite pulp.[23,65] The pulp can then undergo various physical disintegration methods to defibrillate cellulose. These methods are very energy extensive,[19] hence, the energy cost are mitigated by pretreatment of natural fibers and wood pulp. Pre-treatment aims to: (1) limit hydrogen bonding, (2) adding a repulsive charge on the surface of CNF and (3) decreasing the DP.

Carboxylation pretreatment for example, oxidizes cellulose to produce a carboxylated surface which limits inter fibrillar hydrogen bonding in cellulose, sodium ions can also be added to further limit hydrogen bonding.[13] Carboxylation can be performed using TEMPO mediated oxidation.[13] Other treatment methods, such as enzymatic and acidic hydrolysis facilitate defibrillation by reducing the DP of cellulose[66] while acidified-chlorite bleaching aims to delignify natural fibers.[23]

i. TEMPO-mediated Oxidation

TEMPO-mediated oxidation was first employed to selectively oxidize primary hydroxyls in glucans of cellulose.[67] Hydroxyls in C6 of glucopyranose is oxidized to C6 carboxyls, turning it to uronic acids (glucuronic acid).[67] In TEMPO/NaBr/NaClO system, TEMPO-mediated oxidation starts by adding a small amount of TEMPO and NaBr in a solution of polysaccharide suspended in water, its pH is adjusted to between 10 and 11, while oxidation starts as soon as sodium hypochlorite (NaClO) is added.[13] Bleached paper pulps are usually used in TEMPO-mediation oxidation, which yield TEMPO-oxidized cellulose nanofibrils (TOCN), after a mild physical/mechanical disintegration. Besides TEMPO/NaBr/NaClO system, other TEMPO-mediated oxidation type of carboxylations has also been proposed. TEMPO/NaClO/NaClO$_2$ is conducted in acidic pH between pH 3.5 and 6.8.[68] Electromediated oxidation using 4-acetamido-TEMPO has also been performed, in which, nitrosonium salts (from TEMPO) are regenerated by continuous oxidation in anode.[65] TOCN produced in acidic medium yield a higher carboxylate content compare to TOCN produced in alkaline medium.[68]

Carboxylate content was found to be crucial in the disintegration of cellulose to TOCN. A carboxylate content of at least 0.8 mmol/g yielded individualized TOCN of even widths, 5 nm,[68] while an increasing carboxylate content of up to 1.5 mmol/g gave a limit of 3–4 nm[69] in lateral dimension observed under TEM and a height of 3–4 nm[70] observed under AFM.

TEMPO-mediated oxidation has been proven to be an effective pretreatment to facilitate defibrillation of cellulose; however, it has a few drawbacks. TEMPO has been touted as recyclable catalyst due to its high cost; however, it had yet to be demonstrated. Besides, TEMPO is corrosive and its toxicity has yet to be fully characterized.

ii. Acidified Chlorite

Originally, acidified chlorite is used to determine holocellulose content (total cellulose and hemicellulose) by removing lignin in natural fibers.[46] The role of acidified chlorite and the generation of its active species in acid with respect to lignin removal has been discussed under Section 7.2.1 (b). Acidified chlorite pretreatment begins by adding sodium chlorite and acetic acid/acetate buffer to natural fibers suspended in water. Lignin is oxidized while cellulose and hemicellulose remain. Despite that the

Emergence of a New Nanomaterial

treatment is highly selective in degrading lignin, some degree of cellulose, and hemicellulose will be lost due to the acidic environment. The degradation of cellulose and hemicellulose due to acidified chlorite varies from type of biomass.

Biomasses with Klason lignin content between 14.5% and 28.1% were delignified for four consecutive cycles using acidified chlorite and were found to contain only 1.3–2.5% of Klason lignin after delignification.[71] A considerable amount of hemicellulose is removed during pretreatment, as summarized in Table 7.5. CNF can be obtained after defibrillation of holocellulose prepared via acidified chlorite.[16,23] Hemicellulose content was found to be crucial in the defibrillation of holocellulose and to prevent coalescence of microfibrils.[23]

TABLE 7.5 Hemicellulose Lost and Efficiency of Klason Lignin Removal of Several Biomasses after Acidified Chlorite Delignification for Four Consecutive Cycles.

Biomass	Total Cellulose (wt%)	Total Hemicell (wt%)	Klason Lignin (wt%)	Total Cellulose and Hemicell Lost (wt%)	Total Klason Lignin Removed (wt%)
Poplar	42.40	18.90	26.70	1.64	93.63
Corn stover	37.40	26.00	14.50	9.50	84.14
Switchgrass	35.50	25.70	22.80	16.09	89.04
Pine sawdust	45.10	20.70	28.10	32.89	95.37

Source: Adapted from Kumar et al.[71]

iii. Enzymatic Hydrolysis

Enzymatic hydrolysis has also been employed as a pre-treatment to facilitate defibrillation. *Endoglucanase* is used to hydrolyze sulfite pulp, the resulting CNF is homogeneous, has a higher DP than acidic hydrolyzed sulfite pulp.[66]. Besides, enzymatic hydrolysis also prevents blockage in Gaulin homogenizer[66] and Z-shaped microfluidizer.[14]

b) Physical Disintegration

CNF can be prepared immediately from wood pulp or from natural fibers that have undergone pretreatment.

i. Gaulin Homogenizer

The first CNF produced by Turbak et al. uses a very energy intensive process, that is, homogenization using Gaulin homogenizer.[17] Gaulin type homogenizer was initially used to homogenize milk by breaking large fat globules into smaller sizes and dispersing them uniformly throughout the milk. To defibrillate cellulose, cellulose slurries (wood pulp) are pumped through a narrow channel or orifice under high pressure. A combination of large pressure drop and strong shearing forces are created by opening and closing the valve rapidly defibrillates cellulose.[62] This process allows a high degree of defibrillation, however, multiple passes (up to 30 passes) are required to defibrillate Kraft pulp using Gaulin homogenizer alone without any pretreatment.[72] This energy extensive process consumes 71,570 kJ of energy per kg after 20 passes under Gaulin homogenizer.[19]

ii. Microfluidizer

Similarly, micofluidizer uses high pressure to defibrillate cellulose. Intensifier pump generates a high pressure to accelerate cellulose slurries to interaction chamber containing either Z-shaped or Y-shaped channels. The collisions of streams against each other and the channel walls generate a high shear and impact forces.[73] MCC was defibrillated using microfluidizer at pressure of 138 MPa up to 20 passes.[74] Ten to fifteen passes through a microfluidizer defibrillates MCC to large bundles of CNF while 20 passes gave smaller bundles.[74] More than 20 passes, the aspect ratio decreases, indicating that the resulting CNF is damaged.[74] Various unbleached pulps were also defibrillated with only five passes at 55 MPa where the lateral dimensions where found to be 10–30 nm.[73] Microfluidizer is a low energy alternative to defibrillate cellulose while achieving the same degree of defibrillation as Gaulin homogenization.[19]

iii. Grinding

High shearing forces generated by grinding stones can be employed to defibrillate cellulose. Grinders manufactured by Masuko Sangyo composed of one static grindstone and one rotating grindstone revolving at *ca* 1500 RPM.[62] Taniguchi and Okumura first reported using a grinder to produce CNF.[75] CNF obtained from bleached Kraft pulp has a lateral dimension of 20–90 nm after 10 passes through the grinder.[75] Bleached pulp (*Pinus radiata*) was defibrillated using a grinder up to 30 passes.[76] After five passes, CNF with lateral dimensions of 20–50 nm was obtained, further passes did not improve the defibrillation.[76] Another notable study by Abe et al., using pulp with α-cellulose content of 85%, CNF of uniform 15 nm in lateral dimension is obtained after only one pass using grinding.[76] Grinding has been identified as one the defibrillation processes which offers the lowest energy consumption.[19]

iv. High Speed Blender

During agitation (blending), the blades shear and cut at very high speed, up to 37,000 rpm.[16] As the blades spin during high-speed agitation to produce CNF, cellulose defibrillates due to high shearing forces. Besides shearing forces, cavitation bubbles are formed during these high-speed agitations. The implosive collapse of cavitation bubbles emits powerful shockwaves to break its surrounding particles into smaller sizes. TOCN has a high degree of carboxylation, its defibrillation from TEMPO-oxidized cellulose using high speed blender requires a short disintegration time of 4 min.[13,21] Zuluaga et al. studied the defibrillation of banana rachis using a combined alkaline, acidic, and bleaching treatment gave a cellulose bundles of 40–60 nm with occasional microfibrils of 5 nm in lateral dimensions.[77] Uetani and Yano studied the defibrillation holocellulose pulp using a high-speed blender and found that holocellulose CNF has a uniform diameter of 15–20 nm. The CNF produced has a same degree of fibrillation of those produced by grinder. The optimum defibrillation parameters are; 30 min, 0.7 wt% at 37,000 rpm.[16]

v. Ultrasonication

Ultrasonication of frequencies >20 kHz, exerts sound energy to break large particles has been employed extensively to produce nano materials. It has been used extensively in commercial pulp making which promotes formation of fibrils without fiber breakage.[78] Ultrasonication produces a very strong mechanical oscillating power due to cavitation effect which is the driving force in defibrillation of cellulose.[79] A combination of milling and ultrasonication processes toward defibrillation of TEMPO-oxidized bleached Kraft pulp, manage to obtain 96.7% yield of TOCN after 15 min of ultrasonication.[80] The diameters of TOCN obtained falls between 2 nm and 4.5 nm.[80] Wang and Cheng evaluated the defibrillation of four different cellulose sources, the effects of power, time, temperature, concentration, fiber size, and distance of probe to the beaker's bottom. This process gave a mixture of CNF and microscale fibrils in which the highest yield of CNF obtained, accounts for 53% after 60 min of ultrasonication for Avicel. Regenerated cellulose, Lyocell, yielded only 5% after 60 min of ultrasonication. Overall, higher power and higher temperature gave higher yield while concentration and distance of probe have no effect on its yield.[79]

7.2.3 BACTERIAL CELLULOSE (BC)

BC, also known as microbial cellulose or bio-cellulose is produced from acetic acid bacteria, *Acetobacter*.[1] As the name implied, *Acetobacter* produces acetic acid, however, *A. xylinum* bacteria are able to synthesize cellulose which is normally only done by plants.[1] BC extraction differs from CNC or CNF, in which no removal of hemicellulose or lignin is required and no physical agitation is required too.

Besides *Acetobacter*, other bacterial genera such as *Agrobacterium*, *Alcaligenes*, *Pseudomonas*, *Rhizobium,* and *Sarcina* can also produce cellulose.[81] *A. xylinum*, a Gram-negative bacteria is the most profilic producer of BC.[1] For CNC and CNF production, a top-down approach is used, whereas for BC, a bottom-up approach is utilized. Cellulose can be bio-synthesized by a variety of pentoses, hexoses, oligosaccharides, starch, alcohol, and organic acid.[81] The synthesis of BC occurs between the outer and plasma membranes of the cell by cellulose synthesizing complex with surface pores of 3.5 nm in diameter.[82] Cellulose secreted by *A. xylinum* bacteria have a unique ribbon-shaped fibrils of widths of *ca.* 3.5 nm.[81,82]

7.3 NANOCOMPOSITES

Composite material is defined as a material of two or more distinct phases consist of high strength and modulus reinforcing agent bonded to a matrix which retain their original physical and chemical identities yet producing a combination of properties that cannot be achieved with either constituents acting alone.[83] Polymer composites are mixture of polymers with organic or inorganic fillers. Nanocomposites are composite materials with nanofillers in which the reinforcement is in nanoscale, that is, at least one dimension is in nanometer range (<100 nm). Nanocomposites show great promise in terms of superior mechanical, thermal, electrical, and optical properties.[83] Its superior properties can be attributed to the high ratio of surface area to volume which provides a substantial interfacial interaction between reinforcement and matrix.

The recent resurgence of polymer nanocomposites saw a growing need to look beyond formulating polymers with nanofillers but nano-composites that are well engineered, design, and functional.[18] In the past decade, nanofillers such as nanosilicates,[84] carbon nanotubes (CNT)[85], and graphene based fillers[86] are largely investigated. Besides an improvement in mechanical properties, nanofillers such as CNT and graphene imbued matrix with electrical conductivity while nanosilicates gave improvement in thermal stability. In addition, composites require a high amount of filler to reinforce matrix, in contrast, nanocomposites require a low amount of nanofillers.

Bionanocomposites are composites with bio-based nanofillers made from renewable resources with either petroleum-derived polymers (poly-ethylene, polypropylene, polystyrene, etc.) or biopolymers (PLA, poly-hydroxyalkanoates, etc.).[18] Nanocellulose based bionanocomposites is currently in the spotlight due to its low density, renewability, easy avail-ability, biodegradable, relatively reactive surface for chemical modi-fication, or grafting, etc.[18] On the other hand, using nanocellulose as reinforcement in composites present an attractive proposition due to its high strength. CNCs have been researched extensively in recent years and it is ascribed to their high modulus, high stiffness, good aspect-ratio, biode-gradability, biocompatibility, and low density. Previously, a CNC content of 1–5 wt% in PLA did not show an increment of mechanical strength but in turn decreased it[87] while others show a small increment[8]. This is in contrary with composite theory of reinforcement and matrix, where high strength and modulus fibers embedded in a matrix (polymer) reinforces it, provided that the interfacial property which is responsible for its load transfer medium is compatible. The crystallinity of polymer and its effect toward the physical and mechanical properties is well documented where the increase of crystallinity usually translates to better mechanical perfor-mance. Pei et al. pointed out that the increment of mechanical strength was due to increment in crystallinity of PLA caused by the nucleating effect of CNC and better dispersion of modified CNC in PLA.[88] The nucleating effect of CNC and modified CNC in PLA were observed in differential scanning calorimeter and polarized optical microscope after melting at 210C which cannot be compared directly with the non-thermally treated tensile test samples which are solution casted.[88] In another study, a low filler content in composite of CNC (0.1–0.4 wt%) in PLA films were prepared using solution casting method.[89] The optimum modulus increment are, 0.2 wt% CNC in PLA which gave an increment of 125%.[89] It was shown that the resulting improvement of tensile properties were not due to increment of crystallinity in PLA as the resulting composites has a crystallinity which fluctuates between ± 1 and ± 3% with increasing filler content determined using XRD and DSC, respectively.[89] The addition of CNC did not alter the final crystallinity of PLA composites because no thermal treatment was conducted as the films were prepared using solution casting from chloroform.[89]

However, as a reinforcement, CNC and CNF have its disadvantages. Due to its inherently hydrophilic properties (rich in hydroxyls) they tend to absorb moisture, have a poor wettability and compatibility against most of polymer matrices.[18] CNC extracted from sulfuric acidic hydrolysis degrades at a lower temperature, as low as 180°C,[100] whereas for cellulose it is 310°C.[90] Low degradation temperature of CNC limits the type of matrix that can be used in conjunction with CNC as a majority of polymer matrix is thermally processed.

Another downside of CNC and CNF is its difficulty in dispersing well in non-polar medium due to its polar surface.[91] This limits the processing ability and causes dispersion problem in hydrophobic polymers. Currently there a two different techniques used to prepare nanocellulose based bionanocomposites, that is, solution casting of nanocellulose in water or organic solvent and extrusion with freeze-dried nanocelluloses.[18] While, the first technique used for water-soluble polymers, polymer emulsions and other non-water soluble polymers solves some of agglomeration issues of CNC and CNF, the second technique does not. Re-dispersing freeze-dried CNC and CNF in water is difficult due to flocculation and formation of strong hydrogen bonding between them after drying.[7] Hence, chemical modification, polymer grafting, and coating of surface have been conceived to obtain a non-flocculated nanocellulose in non-polar medium (non-polar solvent or hydrophobic polymer).

Emergence of a New Nanomaterial

a) Solution Casting

Solution casting offers a quick, small scale, and non-complex way of obtaining composites. The solution of polymer with nanocellulose is simply evaporated via air-drying or in an oven. CNC obtained after acidic hydrolysis (sulfuric acid) are usually stored in its suspension form. On the other hand, defibrillation of CNF is usually done in water while the final product is suspended in water. Hence, CNC and CNF are both readily available forms of nanocellulose dispersed well in water. However, this method is not practical for industrial scale processing due to limitations of polymer used and time consuming. Nonetheless, it is a suitable for small-scale preparations such as in a laboratory.

i. Water Based

One of the earliest nanocellulose based nanocomposites prepared is tunicate based CNC reinforced latex of copolymer, styrene (35 wt%) and butyl acrylate (65 wt%).[92] Polymers are dispersed in an emulsion of synthetic latex with CNC of weight fraction between 0% and 6% are prepared in a water based system, evaporated at room temperature.[92] It was later examined and described using percolation model by the same group of researchers that CNC form a percolation network due to formation of hydrogen bondings between itself, therefore, reinforcing the composite even at a very low content.[93] CNC reinforced natural rubber (NR) were also prepared and found to effectively reinforced NR as low as 1 wt%.[94] With CNC of weight fraction of <2.5 wt%, a ductile behavior is observed whereas at higher weight fraction, a quasi-brittle behavior is observed.[94] Using dynamic mechanical analysis (DMA), the percolation threshold of CNC with aspect ratio of 43 is determined to be 1.6 v%, equivalent to 2.4 wt%.[94]

Polyvinyl alcohol (PVA), a water soluble and biodegradable polymer, is an ideal polymer to evaluate the reinforcing capability of nanocelluloses.[95] Intermolecular hydrogen bonding can be formed between the rich hydroxyls of nanocellulose and PVA to improve its interfacial properties. Partially hydrolyzed polyvinyl acetate (PVAc) was reinforced by cotton CNC showed an improvement in tensile properties especially for PVAc with higher degree of hydrolysis.[96] PVA reinforced CNF were also found to significantly improve its tensile properties, 55% in modulus with

3.0 wt% of CNF and up to 90% in tensile strength with 5.0 wt% of CNF.[95] Other water soluble polymers such as hydroxypropyl cellulose (HPC) reinforced by TOCN shown an exceptional storage modulus retention at higher temperatures and suppression of matrix glass transition with TOCN *ca.* 5.0 wt%.[80]

ii. Organic Solvent Based

As stated earlier, nanocelluloses prepared as it is are dispersed in a water-based suspension. Using an organic solvent-based solution casting method opens up the option of using largely available hydrophobic polymers. The organic solvent used inadvertently affects the final product due to different polymer solubility, rate of evaporation, and drying conditions.

In order to obtain a stable suspension in organic solvent, surfactant, and polymer grafting can be utilized. CNC is esterified with sulfate groups on its surface after acidic hydrolysis which made it highly negatively charged. Esterified CNC are able to self-assemble, forming a stable chiral nematic phase in polar solvent (water).[5] These chiral nematic phase forms a stable isotropic phase at an onset concentration. Using a commercial surfactant, Beycostat NA (BNA) a phosphoric ester of polyoxyethylene (9) nonylphenyl ether, chiral nematic phase of CNC can also be formed in a non-polar solvent.[97] The same group of researchers then prepared these coated CNC and CNC grafted with maleated polypropylene dispersed in toluene and mixed into solubilized atactic polypropylene (PP) in hot toluene (110°C), heated under vacuum at 90°C for 6 h to remove all solvent follow by hot pressing at 7 MPa for 20 min at 150°C.[98] Both surfactant-coated and grafted CNC dispersed well in PP at 6.0 wt% with occasional aggregates also shown an improvement in tensile properties while surfactant coated CNC showed an improvement in elongation.[98] This shows that a good dispersion of cellulosic nanofillers in a non-polar medium is very important in order to effectively reinforce hydrophobic polymer, for example, PP.

b) Thermal Processing

Nanocellulose-based bionanocomposites can also be prepared using extrusion method.[30] One of the issue of the thermal processing is dispersion, that is, nanocellulose can only be processed in dry state. During the drying process, nanocellulose form aggregates due to strong hydrogen bonds. It was found that CNC prepared via solution casting has a better mechanical property that nanocomposites of the same mixture prepared via freeze drying and hot pressing.[34,99] This is due to formation of percolation network of CNC linked by hydrogen bonds formed by solution casted nanocomposites.

Another challenge in thermal processing is the degradation of CNC during thermal processing. CNC is commonly prepared via hydrolysis of concentrated sulfuric acid, this inadvertently introduce sulfates on the surface of CNC. This causes a significant decrease in degradation temperature (150°C) of cellulose due to sulfates.[100] This can be seen clearly in CNC reinforced polystyrene (PS) in which CNC reinforced PS were prepared using twin-screw extrusion at 200°C, 150 RPM for 10 min.[101] All nanocomposites gradually darken with increasing CNC content, indicating degradation of CNC during extrusion.[101]

7.3.1 BIOMEDICAL APPLICATIONS

Biocompatibility between nanocellulose and human body is of utmost importance if it were to be further utilized for biomedical applications. According to William, the word, "biocompatible" refers to the characterization of a material property which includes a variety of mechanisms of interaction between biomaterials and tissues or tissue components.[102] Hence, biocompatibility is a characteristic not a property, as no material is truly biocompatible except for your own.[102]

It is too early to regard nanocellulose as intrinsically biocompatible as there are not many direct investigations; however, cellulose is considered to be biocompatible while causing only moderate foreign body response.[103] In an *in vivo* bicompatibility study, BC implants caused and benign inflammatory reaction that decreased over time but did not draw foreign body reaction.[104] BC also tends to calcify over a long period of time due to its poor porosity which limits the colonization of material.[104] Hemobiocompatibility of BC as tissue engineered blood vessels were also investigated.[105] Likewise, inflammatory reaction was observed initially, but advanced to a mild chronic inflammation. The inflammation degree were the same even for recombinant protein coated BC, however, BC implants were found to be slightly more irritating compare to expanded polytetrafluoroethylene vascular prosthesis control sample.[105] Modifying the surface properties of nanocellulose may enhance the biocompatibility of nanocellulose. As discussed earlier, CNC surface is mostly esterified with sulfates[10] while CNF were mostly hydroxyls or carboxylates.[65]

No doubt, nanocellulose can potentially serve as next generational biomaterial for biomedical applications. Nanocellulose applications on immobilization of enzymes, tissue scaffolds, tissue replacements, antimicrobial materials, drug excipient, delivery, etc. had been reviewed recently.[103] Some of the applications proposed and published are shown in Table 7.6. Table 7.6 only showed a small fraction of the currently proposed biomedical applications of nanocellulose.

TABLE 7.6 Applications of Nanocellulose in Biomedical Applications.

Applications	Materials	References
Immobilization of enzymes	Peroxidase on CNC	[106]
	Laccase on CNF	[107]
Wound dressing	BC (*Acetobacter xylinum*)	[108]
Dental implants	BC (Gengiflex)	[109]
Bone regeneration	BC (*Acetobacter xylinum*)	[110]
Tissue regeneration	Partially dissolved CNF for ligaments or tendons.	[111]
Tissue scaffold	Dental follicle cells on CNC/cellulose	[112]
Antimicrobial materials	Polyrhodanine coated CNC	[113]
	Silver nanoparticles on BC	[114]
Drug excipient and delivery	Paracetamol in CNF	[115]

7.3.2 ELECTRONICS

Cellulose (paper) is mainly used as packaging material and displaying or storing information. According to Tobjörk et al., printing electronics onto paper starts as early as 1960s, where inorganic thin-film transistors are deposited onto paper substrates in vacuum.[116] Different materials can be imbued onto cellulose to induce different electrical properties, that is, conductivity.[116] Recently, there is a growing interest in using nanocellulose

Emergence of a New Nanomaterial

as a green alternative for flexible electronic substrates. Some of the applications proposed are shown in Table 7.7.

TABLE 7.7 Applications of Nanocellulose in Electronics.

Applications	Materials	References
Solar cell	Organic solar cells on carboxymethylated CNF	[117]
Recyclable solar cell	Organic solar cells on CNC	[118]
Super capacitor	TOCN aerogel as electrode substrate	[119]
Batteries	Li-ion, silicon on carboxymethylated CNF	[120]
Transistor	Organic field-effect transistors on TOCN	[121]
Strain sensor	Graphene-CNF	[122]

7.3.3 BIOADSORBENTS

Cationic dyes are widely used in textiles, cosmetics industries, dyeing and printing.[123] The dye removal from industrial wastewater is necessary due to negative health effects on humans. Besides, the pollution of dye also increases the chemical oxygen demand of water and decreases the projection of lighting for aquatic plants.[124] Dye removal is usually performed using biological treatment, coagulation or flocculation, adsorption, and oxidation.[123] Lignocellulosic materials are gaining interest for its adsorbent applications in wastewater treatment due to its biodegradability and non-toxic nature.[125] Various lignocellulosic and biomass materials have been tested extensively, some of the examples are: wheat shells,[124] peanut hull,[126] algae,[127] wood shavings,[128] kenaf core,[129] and many more.

Cationic dye has a very low affinity toward cellulose.[130] As such, nanocellulose has to rely on hemicellulose and lignin to adsorb dye. Using nanocellulose as bio-adsorbent is quite new. To best of our knowledge, there are several publications on adsorption of nanocelluloses, that is, two for CNC and two for CNF.[131] Both CNC had been shown to be a good adsorbent for cationic dye.[131,132] CNC's maximum adsorption capacity for methylene blue (MB) increases from 118 to 769 mg/g for TEMPO-oxidized CNC. CNC is able to adsorb MB due to negative surface charge of sulfated CNC. By oxidizing CNC using TEMPO-oxidation method, carboxyls functionalities are introduced onto its surface which further increases its adsorption capacity.[132] Surprisingly, aerogel prepared from TEMPO-oxidized CNF has a low maximum adsorption capacity of 3.70 mg/g toward MB.[133] However, another study had shown that CNF has a maximum adsorption capacity of up to 122.2 mg/g, in which CNF richer in hemicellulose content has a higher adsorption capacity than CNF of lower hemicellulose content.[134]

7.3.4 RHEOLOGY MODIFIER

The high aspect ratio of CNF binds to water strongly while retaining its solid structure. The result is a gel-like substance in which water fills most of its volume.[25] CNF "gel" gives unique rheological properties; even at low nanocellulose concentrations it gives high viscosity. CNC and CNF at 3% concentration have a high storage modulus of 104 and 102 Pa, respectively.[14,135] Nanocellulose gels also exhibits shear-thinning behavior, that is, the viscosity decreases with shear forces (flow).[25] Such behavior observed coupled with water retention may be beneficial for applications in coating (paint) and low calorie food thickening agent.[25]

7.4 FUTURE PERSPECTIVES

This chapter summaries processes for the production of nanocelluloses and their applications. Recently, nanocelluloses, especially CNC and CNF, have been commercially produced in large scale, in which the production is expected to be increased remarkably in near future. Besides research on the isolation of nanocelluloses and other potential applications, characterizations of these new nanomaterials are also critically important.

7.5 ACKNOWLEDGMENTS

The authors would like to thank the Universiti Kebangsaan Malaysia and the Ministry of Higher Education for providing research funding DIP-2014-013 and LRGS/TD/2012/USM-UKM/PT/04, respectively.

KEYWORDS

- **extraction methods**
- **nanocellulose**
- **applications**

REFERENCES

1. Klemm, D.; Kramer, F.; Moritz, S.; Lindström, T.; Ankerfors, M.; Gray, D.; Dorris, A. Nanocelluloses: A New Family of Nature-Based Materials. *Angew. Chem. Int. Ed.* **2011,** *50,* 5438–5466.
2. Gibson, L. J. The Hierarchical Structure and Mechanics of Plant Materials. *J. R. Soc. Interface.* **2012,** *9,* 2749–2766.
3. Iwamoto, S.; Kai, W.; Isogai, A.; Iwata, T. Elastic Modulus of Single Cellulose Microfibrils from Tunicate Measured by Atomic Force Microscopy. *Biomacromolecules.* **2009,** *10,* 2571–2576.
4. Mukherjee, S. M.; Woods, H. J. X-ray and Electron Microscope Studies of The Degradation of Cellulose by Sulphuric Acid. *Biochimica. Et. Biophysica. Acta.* **1953,** *10,* 499–511.
5. Dong, X. M.; Revol, J. F.; Gray, D. G. Effect of Microcrystallite Preparation Conditions on The Formation of Colloid Crystals of Cellulose. *Cellulose.* **1998,** *5,* 19–32.
6. Beck, S.; Bouchard, J.; Berry, R. Controlling the Reflection Wavelength of Iridescent Solid Films of Nanocrystalline Cellulose. *Biomacromolecules.* **2010,** *12,* 167–172.
7. Habibi, Y.; Lucia, L. A.; Rojas, O. J. Cellulose Nanocrystals: Chemistry, Self-Assembly, and Applications. *Chem. Rev.* **2010,** *110,* 3479–3500.
8. Bondeson, D.; Oksman, K. Polylactic Acid/Cellulose Whisker Nanocomposites Modified by Polyvinyl Alcohol. *Compos. Part A: Appl. Sci. Manuf.* **2007,** *38,* 2486–2492.
9. Elazzouzi-Hafraoui, S.; Nishiyama, Y.; Putaux, J. L.; Heux, L.; Dubreuil, F.; Rochas, C. The Shape and Size Distribution of Crystalline Nanoparticles Prepared by Acid Hydrolysis of Native Cellulose. *Biomacromolecules.* **2007,** *9,* 57–65.
10. Dong, X. M.; Kimura, T.; Revol, J. F.; Gray, D. G. Effects of Ionic Strength on The Isotropic - Chiral Nematic Phase Transition of Suspensions of Cellulose Crystallites. *Langmuir.* **1996,** *12,* 2076–2082.
11. Dong, X. M.; Kimura, T.; Revol, J. F.; Gray, D. G. Effects of Ionic Strength on the Isotropic-Chiral Nematic Phase Transition of Suspensions of Cellulose Crystallites. *Langmuir.* **1996,** *12,* 2076–2082.
12. Cheung, C. C. Y.; Giese, M.; Kelly, J. A.; Hamad, W. Y.; MacLachlan, M. J. Iridescent Chiral Nematic Cellulose Nanocrystal/Polymer Composites Assembled in Organic Solvents. *ACS Macro Lett.* **2013,** *2,* 1016–1020.

13. Isogai, A.; Saito, T.; Fukuzumi, H. TEMPO-Oxidized Cellulose Nanofibers. *Nanoscale.* **2011,** *3,* 71–85.
14. Pääkkö, M.; Ankerfors, M.; Kosonen, H.; Nykänen, A.; Ahola, S.; Österberg, M.; Ruokolainen, J.; Laine, J.; Larsson, P. T.; Ikkala, O.; Lindström, T. Enzymatic Hydrolysis Combined with Mechanical Shearing and High-Pressure Homogenization for Nanoscale Cellulose Fibrils and Strong Gels. *Biomacromolecules.* **2007,** *8,* 1934–1941.
15. Deepa, B.; Abraham, E.; Cherian, B. M.; Bismarck, A.; Blaker, J. J.; Pothan, L. A.; Leao, A. L.; de Souza, S. F.; Kottaisamy, M. Structure, Morphology and Thermal Characteristics of Banana Nano Fibers Obtained by Steam Explosion. *Bioresour. Technol.* **2011,** *102,* 1988–1997.
16. Uetani, K.; Yano, H. Nanofibrillation of Wood Pulp Using a High-Speed Blender. *Biomacromolecules.* **2010,** *12,* 348–353.
17. Turbak, A. F.; Snyder, F. W.; Sandberg, K. R. In *Microfibrillated Cellulose, a New Cellulose Product: Properties, Uses, and Commercial Potential.* Proceeding of the Ninth Cellulose Conference. Applied Polymer Symposia 37; Sarko, A., Ed.; Wiley: New York City, 1983; pp 815–827.
18. Siqueira, G.; Bras, J.; Dufresne, A. Cellulosic Bionanocomposites: A Review of Preparation, Properties and Applications. *Polymers.* **2010,** *2,* 728–765.
19. Spence, K.; Venditti, R.; Rojas, O.; Habibi, Y.; Pawlak, J. A Comparative Study of Energy Consumption and Physical Properties of Microfibrillated Cellulose Produced by Different Processing Methods. *Cellulose.* **2011,** *18,* 1097–1111.
20. Lavoine, N.; Desloges, I.; Dufresne, A.; Bras, J. Microfibrillated Cellulose – Its Barrier Properties and Applications in Cellulosic Materials: A Review. *Carbohydr. Polym.* **2012,** *90,* 735–764.
21. Saito, T.; Nishiyama, Y.; Putaux, J. L.; Vignon, M.; Isogai, A. Homogeneous Suspensions of Individualized Microfibrils From TEMPO-Catalyzed Oxidation of Native Cellulose. *Biomacromolecules.* **2006,** *7,* 1687–1691.
22. Parikka, K.; Leppänen, A. S.; Xu, C.; Pitkänen, L.; Eronen, P.; Österberg, M.; Brumer, H.; Willför, S.; Tenkanen, M. Functional and Anionic Cellulose-Interacting Polymers by Selective Chemo-Enzymatic Carboxylation of Galactose-Containing Polysaccharides. *Biomacromolecules.* **2012,** *13,* 2418–2428.
23. Iwamoto, S.; Abe, K.; Yano, H. The Effect of Hemicelluloses on Wood Pulp Nanofibrillation and Nanofiber Network Characteristics. *Biomacromolecules.* **2008,** *9,* 1022–1026.
24. Cheng, Q.; Wang, S.; Rials, T. G. Poly(Vinyl Alcohol) Nanocomposites Reinforced with Cellulose Fibrils Isolated by High Intensity Ultrasonication. *Compos.Part A: Appl. Sci. Manuf.* **2009,** *40,* 218–224.
25. Dimic-Misic, K.; Gane, P. A. C.; Paltakari, J. Micro- and Nanofibrillated Cellulose as a Rheology Modifier Additive in CMC-Containing Pigment-Coating Formulations. *Ind. Eng. Chem. Res.* **2013,** *52,* 16066–16083.
26. Zheng, G.; Cui, Y.; Karabulut, E.; Wågberg, L.; Zhu, H.; Hu, L. Nanostructured Paper for Flexible Energy and Electronic Devices. *MRS Bull.* **2013,** *38,* 320–325.
27. Ninan, N.; Muthiah, M.; Park, I. K.; Elain, A.; Thomas, S.; Grohens, Y. Pectin/ Carboxymethyl Cellulose/Microfibrillated Cellulose Composite Scaffolds for Tissue Engineering. *Carbohydr. Polymers.* **2013,** *98,* 877–885.

28. Ribi, E. Submicroscopic Structure of Fibres nd Their Formation. *Nature.* **1951,** *168,* 1082–1083.
29. Petersson, L.; Kvien, I.; Oksman, K. Structure and Thermal Properties of Poly(Lactic Acid)/Cellulose Whiskers Nanocomposite Materials. *Compos. Sci. Techno.* **2007,** *67,* 2535–2544.
30. Oksman, K.; Mathew, A. P.; Bondeson, D.; Kvien, I. Manufacturing Process of Cellulose Whiskers/Polylactic Acid nanocomposites. *Compos. Sci. Technol.* **2006,** *66,* 2776–2784.
31. Paralikar, S. A.; Simonsen, J.; Lombardi, J. Poly(vinyl alcohol)/Cellulose Nanocrystal Barrier Membranes. *J. Memb. Sci.* **2008,** *320,* 248–258.
32. Grunert, M.; Winter, W. Nanocomposites of cellulose acetate butyrate reinforced with cellulose nanocrystals. *J. Polym. Environ.* **2002,** *10,* 27–30.
33. Morandi, G.; Heath, L.; Thielemans, W. Cellulose Nanocrystals Grafted with Polystyrene Chains through Surface-Initiated Atom Transfer Radical Polymerization (SI-ATRP). *Langmuir.* **2009,** *25,* 8280–8286.
34. Dufresne, A.; Cavaillé, J. Y.; Helbert, W. Thermoplastic Nanocomposites Filled with Wheat Straw Cellulose Whiskers. Part II: Effect of Processing and Modeling. *Polymer Compos.* **1997,** *18,* 198–210.
35. Alemdar, A.; Sain, M. Biocomposites From Wheat Straw Nanofibers: Morphology, Thermal and Mechanical Properties. *Compos. Sci. Technol.* **2008,** *68,* 557–565.
36. Andresen, M.; Stenius, P. Water-in-Oil Emulsions Stabilized by Hydrophobized Microfibrillated Cellulose. *J. Dispers. Sci. Technol.* **2007,** *28,* 837–844.
37. Dinand, E.; Chanzy, H.; Vignon, R. M. Suspensions of Cellulose Microfibrils From Sugar Beet Pulp. *Food Hydrocoll.* **1999,** *13,* 275–283.
38. Mörseburg, K.; Chinga-Carrasco, G. Assessing the Combined Benefits of Clay and Nanofibrillated Cellulose in Layered TMP-Based Sheets. *Cellulose.* **2009,** *16,* 795–806.
39. Morán, J.; Alvarez, V.; Cyras, V.; Vázquez, A. Extraction of Cellulose and Preparation of Nanocellulose From Sisal Fibers. *Cellulose.* **2008,** *15,* 149–159.
40. Cheng, H.; Zhan, H.; Fu, S.; Lucia, L. A. Alkali Extraction of Stover. *Bioresources.* **2010,** *11,* 196–206.
41. Lundquist, K.; Simonson, R.; Tingsvik, K. Lignin-Carbohydrate Linkages in Milled-Wood Lignin Preparation. *Svensk Papperstidn.* **1983,** *86,* 44–47.
42. Yoon, S. Y.; Han, S. H.; Shin, S. J. The Effect of Hemicelluloses and Lignin on Acid Hydrolysis of Cellulose. *Energy.* **2014,** *77,* 19–24.
43. Sixta, H.; Süss, H. U.; Potthast, A.; Schwanninger, M.; Krotscheck, A. W. Pulp Bleaching. In *Handbook of Pulp*; Sixta, H., Ed.; Wiley-VCH Verlag GmbH: Weinheim, 2008.
44. Lachenal, D.; Muguet, M. Degradation of Residual Lignin in Kraft Pulp with Ozone. Application to Bleaching. *Nord. Pulp Pap. Res. J.* **1992,** *7,* 25–29.
45. Liu, J.; Zhou, X. F. Structural Changes in Residual Lignin of *Eucalyptus Urophylla* Eucalyptus Grandis LH 107 Oxygen Delignified Kraft Pulp Upon Chlorine Dioxide Bleaching. *Sci. Iran.* **2011,** *18,* 486–490.
46. Wise, L. E.; Murphy, M.; D'Addieco, A. A. Chlorite Holocellulose, Its Fractionation and Bearing on Summative Wood Analysis and on Studies on the Hemicelluloses. *Pap. Trade J.* **1946,** *122,* 35–43.

47. Yu, H.; Qin, Z.; Liang, B.; Liu, N.; Zhou, Z.; Chen, L. Facile Extraction of Thermally Stable Cellulose Nanocrystals with a High Yield of 93% Through Hydrochloric Acid Hydrolysis Under Hydrothermal Conditions. *J. Mater. Chem. A.* **2013,** *1,* 3938–3944.

48. Sadeghifar, H.; Filpponen, I.; Clarke, S.; Brougham, D.; Argyropoulos, D. Production of Cellulose Nanocrystals Using Hydrobromic Acid and Click Reactions on Their Surface. *J. Mater. Sci.* **2011,** *46,* 7344–7355.

49. Zhang, P. P.; Tong, D. S.; Lin, C. X.; Yang, H. M.; Zhong, Z. K.; Yu, W. H.; Wang, H.; Zhou, C. H. Effects of Acid Treatments on Bamboo Cellulose Nanocrystals. *Asia-Pacific J. Chem. Eng.* **2014,** *9,* 686–695.

50. Dufresne, A. Processing of Polymer Nanocomposites Reinforced with Polysaccharide Nanocrystals. *Molecules.* **2010,** *15,* 4111–4128.

51. Araki, J.; Wada, M.; Kuga, S.; Okano, T. Influence of Surface Charge on Viscosity Behavior of Cellulose Microcrystal Suspension. *J. Wood Sci.* **1999,** *45,* 258–261.

52. Kargarzadeh, H.; Ahmad, I.; Abdullah, I.; Dufresne, A.; Zainudin, S.; Sheltami, R. Effects of Hydrolysis Conditions on The Morphology, Crystallinity, and Thermal Stability of Cellulose Nanocrystals Extracted From Kenaf Bast Fibers. *Cellulose.* **2012,** *19,* 855–866.

53. Chan, C. H.; Chia, C. H.; Zakaria, S.; Ahmad, I.; Dufresne, A. Production and Characterisation of Cellulose and Nano-Crystalline Cellulose From Kenaf Core Wood. *Bio. Resour.* **2013,** *8,* 785–794.

54. Revol, J. F. On the Cross-Sectional Shape of Cellulose Crystallites in Valonia Ventricosa. *Carbohydr. Polym.* **1982,** *2,* 123–134.

55. Marchessault, R. H.; Morehead, F. F.; Walter, N. M. Liquid Crystal Systems From Fibrillar Polysaccharides. *Nature.* **1959,** *184,* 632–633.

56. Millet, M. A.; Moore, W. E.; Saeman, J. E. Preparation and Properties of Hydrocelluloses. *Ind. Eng. Chem.* **1954,** *46,* 1493–1497.

57. Eichhorn, S.; Dufresne, A.; Aranguren, M.; Marcovich, N.; Capadona, J.; Rowan, S.; Weder, C.; Thielemans, W.; Roman, M.; Renneckar, S.; Gindl, W.; Veigel, S.; Keckes, J.; Yano, H.; Abe, K.; Nogi, M.; Nakagaito, A.; Mangalam, A.; Simonsen, J.; Benight, A.; Bismarck, A.; Berglund, L.; Peijs, T. Review: Current International Research into Cellulose Nanofibres and Nanocomposites. *J. Mater. Sci.* **2010,** *45,* 1–33.

58. Kovacs, T.; Naish, V.; O'Connor, B.; Blaise, C.; Gagné, F.; Hall, L.; Trudeau, V.; Martel, P. An Ecotoxicological Characterization of Nanocrystalline Cellulose (NCC). *Nanotoxicology.* **2010,** *4,* 255–270.

59. Babacar, L. E.; Bras, J.; Sadocco, P.; Belgacem, M. N.; Dufresne, A.; Thielemans, W. Surface Functionalization of Cellulose by Grafting Oligoether Chains. *Mater. Chem. Phys.* **2010,** *120,* 438–445.

60. Lin, N.; Dufresne, A. Surface Chemistry, Morphological Analysis and Properties of Cellulose Nanocrystals with Gradiented Sulfation Degrees. *Nanoscale.* **2014,** *6,* 5384–5393.

61. Manley, R. S. J. Fine Structure of Native Cellulose Microfibrils. *Nature.* **1964,** *204,* 1155–1157.

62. Plackett, D.; Iotti, M. Preparation of Nanofibrillated Cellulose and Cellulose Whiskers. In *Biopolymer Nanocomposites;* John Wiley & Sons, Inc.: Hoboken, NJ, 2013; pp 309–338.

63. Jiang, F.; Hsieh, Y. L. Super Water Absorbing and Shape Memory Nanocellulose Aerogels From TEMPO-Oxidized Cellulose Nanofibrils via Cyclic Freezing-Thawing. *J. Mater. Chem. A.* **2014,** *2,* 350–359.

64. Li, Q.; Renneckar, S. Supramolecular Structure Characterization of Molecularly Thin Cellulose I Nanoparticles. *Biomacromolecules.* **2011,** *12,* 650–659.

65. Isogai, T.; Saito, T.; Isogai, A. TEMPO Electromediated Oxidation of Some Polysaccharides Including Regenerated Cellulose Fiber. *Biomacromolecules.* **2010,** *11,* 1593–1599.

66. Henriksson, M.; Henriksson, G.; Berglund, L. A.; Lindström, T. An Environmentally Friendly Method for Enzyme-Assisted Preparation of Microfibrillated Cellulose (MFC) Nanofibers. *Eur. Polym. J.* **2007,** *43,* 3434–3441.

67. de Nooy, A. E. J.; Besemer, A. C.; van Bekkum, H. Highly Selective Nitroxyl Radical-Mediated Oxidation of Primary Alcohol Groups in Water-Soluble Glucans. *Carbohydr. Res.* **1995,** *269,* 89–98.

68. Hirota, M.; Tamura, N.; Saito, T.; Isogai, A. Oxidation of Regenerated Cellulose with NaClO2 Catalyzed by TEMPO and NaClO Under Acid-Neutral Conditions. *Carbohydr. Polym.* **2009,** *78,* 330–335.

69. Saito, T.; Kimura, S.; Nishiyama, Y.; Isogai, A. Cellulose Nanofibers Prepared by TEMPO-Mediated Oxidation of Native Cellulose. *Biomacromolecules.* **2007,** *8,* 2485–2491.

70. Fukuzumi, H.; Saito, T.; Iwata, T.; Kumamoto, Y.; Isogai, A. Transparent and High Gas Barrier Films of Cellulose Nanofibers Prepared by TEMPO-Mediated Oxidation. *Biomacromolecules.* **2008,** *10,* 162–165.

71. Kumar, R.; Hu, F.; Hubbell, C. A.; Ragauskas, A. J.; Wyman, C. E. Comparison of Laboratory Delignification Methods, their Selectivity, and Impacts on Physiochemical Characteristics of Cellulosic Biomass. *Bioresour. Technol.* **2013,** *130,* 372–381.

72. Nakagaito, A. N.; Yano, H. The Effect of Morphological Changes from Pulp Fiber Towards Nano-Scale Fibrillated Cellulose on the Mechanical Properties of High-Strength Plant Fiber Based Composites. *Appl. Phys. A.* **2004,** *78,* 547–552.

73. Ferrer, A.; Filpponen, I.; Rodríguez, A.; Laine, J.; Rojas, O. J. Valorization of Residual Empty Palm Fruit Bunch Fibers (EPFBF) by Microfluidization: Production of Nanofibrillated Cellulose and EPFBF Nanopaper. *Bioresour. Technol.* **2012,** *125,* 249–255.

74. Lee, S. Y.; Chun, S. J.; Kang, I. A.; Park, J. Y. Preparation of Cellulose Nanofibrils by High-Pressure Homogenizer and Cellulose-Based Composite Films. *J. Ind. Eng. Chem.* **2009,** *15,* 50–55.

75. Taniguchi, T.; Okamura, K. New Films Produced From Microfibrillated Natural Fibres. *Polym. Int.* **1998,** *47,* 291–294.

76. Abe, K.; Iwamoto, S.; Yano, H. Obtaining Cellulose Nanofibers with a Uniform Width of 15 nm from Wood. *Biomacromolecules.* **2007,** *8,* 3276–3278.

77. Zuluaga, R.; Putaux, J. L.; Cruz, J.; Vélez, J.; Mondragon, I.; Gañán, P. Cellulose Microfibrils From Banana Rachis: Effect of Alkaline Treatments on Structural and Morphological Features. *Carbohydr. Polym.* **2009,** *76,* 51–59.

78. Thompson, A. M. R. The Influence of Ultrasound on Virgin Paper Fibers. *Prog. Paper Recycl.* **2002,** *11,* 6–12.

79. Wang, S.; Cheng, Q. A Novel Process to Isolate Fibrils from Cellulose Fibers by High-Intensity Ultrasonication, Part 1: Process Optimization. *J. Appl. Polym. Sci.* **2009,** *113,* 1270–1275.

80. Johnson, R.; Zink-Sharp, A.; Renneckar, S.; Glasser, W. A New Bio-Based Nanocomposite: Fibrillated TEMPO-Oxidized Celluloses in Hydroxypropylcellulose Matrix. *Cellulose.* **2009,** *16,* 227–238.

81. El-Saied, H.; Basta, A. H.; Gobran, R. H. Research Progress in Friendly Environmental Technology for the Production of Cellulose Products (Bacterial Cellulose and its Application). *Polym. Plast. Technol. Eng,* **2004,** *43,* 797–820.

82. Klemm, D.; Heublein, B.; Fink, H.-P.; Bohn, A. Cellulose: Fascinating Biopolymer and Sustainable Raw Material. *Angew. Chem. Int. Ed.* **2005,** *44,* 3358–3393.

83. Mallick, P. K. *Fiber-Reinforced Composites: Materials, Manufacturing, and Design.* CRC Press: Boca Raton, London, 2008.

84. Alexandre, M.; Dubois, P. Polymer-Layered Silicate Nanocomposites: Preparation, Properties and Uses of a New Class of Materials. *Mater. Sci. Eng. R: Reports.* **2000,** *28,* 1–63.

85. Thostenson, E. T.; Ren, Z.; Chou, T.-W. Advances in the Science and Technology of Carbon Nanotubes and their Composites: A Review. *Compos. Sci. Technol.* **2001,** *61,* 1899–1912.

86. Huang, X.; Qi, X.; Boey, F.; Zhang, H. Graphene-Based Composites. *Chem. Soc. Rev.* **2012,** *41,* 666–686.

87. Sanchez-Garcia, M.; Lagaron, J. On the Use of Plant Cellulose Nanowhiskers to Enhance the Barrier Properties of Polylactic Acid. *Cellulose.* **2010,** *17,* 987–1004.

88. Pei, A.; Zhou, Q.; Berglund, L. A. Functionalized Cellulose Nanocrystals as Biobased Nucleation Agents in Poly (L-Lactide)(PLLA)–Crystallization and Mechanical Property Effects. *Compos. Sci. Technol.* **2010,** *70,* 815–821.

89. Chan, C. H.; Chia, C.H.; Zakaria, S.; Ahmad, I.; Dufresne, A.; Tshai, K.Y. Low Filler Content Cellulose Nanocrystal and Graphene Oxide Reinforced Polylactic Acid Film Composites. *Polym. Res. J.* **2014,** *9,* 165–177.

90. Yang, H.; Yan, R.; Chen, H.; Lee, D. H.; Zheng, C. Characteristics of Hemicellulose, Cellulose and Lignin Pyrolysis. *Fuel.* **2007,** *86,* 1781–1788.

91. Khalil, H. P. S.; Davoudpour, Y.; Islam, M. N.; Mustapha, A.; Sudesh, K.; Dungani, R.; Jawaid, M. Production and Modification of Nanofibrillated Cellulose Using Various Mechanical Processes: A Review. *Carbohydr. Polym.* **2014,** *99,* 649–665.

92. Favier, V.; Canova, G. R.; Cavaillé, J. Y.; Chanzy, H.; Dufresne, A.; Gauthier, C. Nanocomposite Materials from Latex and Cellulose Whiskers. *Polym. Adv. Technol.* **1995,** *6,* 351–355.

93. Favier, V.; Cavaille, J. Y.; Canova, G. R.; Shrivastava, S. C. Mechanical Percolation in Cellulose Whisker Nanocomposites. *Polym. Eng. Sci.* **1997,** *37,* 1732–1739.

94. Bendahou, A.; Habibi, Y.; Kaddami, H.; Dufresne, A. Physico-Chemical Characterization of Palm from Phoenix Dactylifera-L, Preparation of Cellulose Whiskers and Natural Rubber-Based Nanocomposites. *JBMB.* **2009,** *3,* 81–90.

95. Frone, A. N.; Panaitescu, D. M.; Donescu, D.; Spataru, C. I.; Radovici, C.; Trusca, R.; Somoghi, R. Preparation and Characterization of PVA Composites with Cellulose Nanofibers Obtained by Ultrasonication. *Bioresources.* **2011,** *6,* 487–512.

96. Roohani, M.; Habibi, Y.; Belgacem, N. M.; Ebrahim, G.; Karimi, A. N.; Dufresne, A. Cellulose Whiskers Reinforced Polyvinyl Alcohol Copolymers Nanocomposites. *Eur. Polym. J.* **2008,** *44,* 2489–2498.
97. Heux, L.; Chauve, G.; Bonini, C. Nonflocculating and Chiral-Nematic Self-Ordering of Cellulose Microcrystals Suspensions in Nonpolar Solvents. *Langmuir.* **2000,** *16,* 8210–8212.
98. Ljungberg, N.; Bonini, C.; Bortolussi, F.; Boisson, C.; Heux, L. Cavaillé New Nanocomposite Materials Reinforced with Cellulose Whiskers in Atactic Polypropylene: Effect of Surface and Dispersion Characteristics. *Biomacromolecules.* **2005,** *6,* 2732–2739.
99. Helbert, W.; Cavaillé, J. Y.; Dufresne, A. Thermoplastic Nanocomposites Filled with Wheat Straw Cellulose Whiskers. Part I: Processing and Mechanical Behavior. *Polym. Compos.* **1996,** *17,* 604–611.
100. Roman, M.; Winter, W. T. Effect of Sulfate Groups from Sulfuric Acid Hydrolysis on the Thermal Degradation Behavior of Bacterial Cellulose. *Biomacromolecules.* **2004,** *5,* 1671–1677.
101. Lin, N.; Dufresne, A. Physical and/or Chemical Compatibilization of Extruded Cellulose Nanocrystal Reinforced Polystyrene Nanocomposites. *Macromolecules.* **2013,** *46,* 5570–5583.
102. Williams, D. F. There is no Such Thing as a Biocompstible Material. *Biomaterials.* **2014,** *35,* 10009–10014.
103. Lin, N.; Dufresne, A. Nanocellulose in Biomedicine: Current Status and Future Prospect. *Eur. Polym. J.* **2014,** *59,* 302–325.
104. Pértile, R. A.; Moreira, S.; Costa, R. M.; Correia, A.; Guardão, L.; Gartner, F.; Vilanova, M. M. G. Bacterial Cellulose: Long-Term Biocompatibility Studies. *J. Biomater. Sci.* **2011,** *23,* 1339–1354.
105. Andrade, F. K.; Alexandre, N.; Amorim, I.; Gartner, F.; Mauricio, A. C.; Luis, A. L.; Gama, M. Studies on the Biocompatibility of Bacterial Cellulose. *J. Bioact. Compat. Polym. Biomed. Appl.* **2012,** *28,* 97–112.
106. Yang, R.; Tan, H.; Wei, F.; Wang, S. Peroxidase Conjugate of Cellulose Nanocrystals for the Removal of Chlorinated Phenolic Compounds in Aqueous Solution. *Biotechnology.* **2008,** *7,* 233–241.
107. Sathishkumar, P.; Kamala-Kannan, S.; Cho, M.; Kim, J. S.; Hadibarata, T.; Salim, M. R.; Oh, B.-T. Laccase Immobilization on Cellulose Nanofiber: The Catalytic Efficiency and Recyclic Application for Simulated Dye Effluent Treatment. *J. Mol. Catal. B: Enzym.* **2014,** *100,* 111–120.
108. Fontana, J. D.; De Souza, A. M.; Fontana, C. K.; Torriani, I. L.; Moreschi, J. C.; Gallotti, B. J.; De Souza, S. J.; Narcisco, G. P.; Bichara, J. A.; Farah, L. F. X. Acetobacter Cellulose Pellicle as a Temporary Skin Substitute. *Appl. Biochem. Biotechnol.* **1990,** *24–25,* 253–264.
109. Novaes Jr, A. B.; Novaes, A. B.; Grisi, M. F. M.; Soares, U. N.; Gabarra, F. Gengiflex, an Alkali-Cellulose Membrane for GTR: Histologic Observations. *Braz. Dent. J.* **1993,** *4,* 65–71.
110. Chen, Y. M.; Xi, T.; Zheng, Y. In Vitro Cytotoxicity of Bacterial Cellulose Scaffolds Used for Tissue-Engineered Bone. *J. Bioact. Compat. Polym. Biomed. Appl.* **2009,** *24,* 137–145.

194   Functional Polymeric Composites: Macro to Nanoscales

111. Mathew, A. P.; Oksman, K.; Pierron, D.; Harmand, M.-F. Fibrous Cellulose Nanocomposite Scaffolds Prepared by Partial Dissolution for Potential Use as Ligament or Tendon Substitutes. *Carbohydr. Polym.* **2012,** *87,* 2291–2298.

112. He, X.; Xiao, Q.; Lu, C.; Wang, Y.; Zhang, X.; Zhao, J.; Zhang, W.; Zhang, X.; Deng, Y. Uniaxially Aligned Electrospun all-Cellulose Nanocomposite Nanofibers Reinforced with Cellulose Nanocrystals: Scaffold for Tissue Engineering. *Biomacromolecules.* **2014,** *15,* 618–627.

113. Tang, J.; Song, Y.; Tanvir, S.; Anderson, W. A.; Berry, R. M.; Tam, K. C. Polyrhodanine Coated Cellulose Nanocrystals: A Sustainable Antimicrobial Agent. *ACS Sustain. Chem. Eng.* **2015,** *3,* 1801–1809.

114. Ifuku, S.; Tsuji, M.; Morimoto, M.; Saimoto, H.; Yano, H. Synthesis of Silver Nanoparticles Templated by TEMPO-Mediated Oxidized Bacterial Cellulose Nanofibers. *Biomacromolecules.* **2009,** *10,* 2714–2717.

115. Kolakovic, R.; Peltonen, L.; Laaksonen, T.; Putkisto, K.; Laukkanen, A.; Hirnoven, J. Spray-Dried Cellulose Nanofibers as Novel Tablet Excipient. *AAPS Pharm. Sci. Tech.* **2011,** *12,* 1366–1373.

116. Tobjörk, D.; Österbacka, R. Paper Electronics. *Adv. Mater.* **2011,** *23,* 1935–1961.

117. Hu, L.; Zheng, G.; Yao, J.; Liu, N.; Weil, B.; Eskilsson, M.; Karabulut, E.; Ruan, Z.; Fan, S.; Bloking, J. T.; McGehee, M. D.; Wågberg, L.; Cui, Y. Transparent and Conductive Paper from Nanocellulose Fibers. *Energy Environ. Sci.* **2013,** *6,* 513–518.

118. Zhou, Y.; Fuentes-Hernandez, C.; Khan, T. M.; Liu, J.-C.; Hsu, J.; Shim, J. W.; Dindar, A.; Youngblood, J. P.; Moon, R. J.; Kippelen, B. Recyclable Organic Solar Cells on Cellulose Nanocrystal Substrates. *Sci. Rep.* **2013,** *3,* 1536.

119. Nyström, G.; Marais, A.; Karabulut, E.; Wågberg, L.; Cui, Y.; Hamedi, M. M. Self-Assembled Three-Dimensional and Compressible Interdigitated Thin-Film Supercapacitors and Batteries. *Nat. Commun.* **2015,** *6,* 7259.

120. Hu, L.; Liu, N.; Eskilsson, M.; Zheng, G.; McDonough, J.; Wågberg, L.; Cui, Y. Silicon-Conductive Nanopaper for Li-Ion Batteries. *Nano Energy.* **2013,** *2,* 138–145.

121. Huang, J.; Zhu, H.; Chen, Y.; Preston, C.; Rohrbach, K.; Cumings, J.; Hu, L. Highly Transparent and Flexible Nanopaper Transistors. *ACS Nano.* **2013,** *7,* 2106–2133.

122. Yan, C.; Wang, J.; Kang, W.; Cui, M.; Wang, X.; Foo, C. Y.; Chee, K. J.; Lee, P. S. Highly Stretchable Piezoresistive Graphene-Nanocellulose Nanopaper for Strain Sensors. *Adv. Mater.* **2013,** *26,* 2022–2027.

123. Dragan, E. S.; Apopei, D. F. Synthesis and Swelling Behavior of Ph-Sensitive Semi-Interpenetrating Polymer Network Composite Hydrogels Based on Native and Modified Potatoes Starch as Potential Sorbent for Cationic Dyes. *Chem. Eng. J.* **2011,** *178,* 252–263.

124. Bulut, Y.; Aydin, H. A. Kinetics and Thermodynamics Study of Methylene Blue Adsorption on Wheat Shells. *Desalination.* **2006,** *194,* 259–267.

125. Crini, G.; Badot, P.-M. Application of Chitosan, a Natural Aminopolysaccharide, for Dye Removal from Aqueous Solutions by Adsorption Processes Using Batch Studies: A Review of Recent Literature. *Prog. Polym. Sci.* **2008,** *33,* 399–447.

126. Gong, R.; Sun, Y.; Chen, J.; Liu, H.; Yang, C. Effect of Chemical Modification on Dye Adsorption Capacity of Peanut Hull. *Dyes Pigments.* **2005,** *67,* 175–181.

127. Vilar, V. J. P.; Botelho, C. M. S.; Boaventura, R. A. R. Methylene Blue Adsorption by Algal Biomass Based Materials: Biosorbents Characterization and Process Behaviour. *J. Hazard. Mater.* **2007,** *147,* 120–132.

128. Janoš, P.; Coskun, S.; Pilařová, V.; Rejnek, J. Removal of Basic (Methylene Blue) and Acid (Egacid Orange) Dyes from Waters by Sorption on Chemically Treated Wood Shavings. *Bioresour. Technol.* **2009,** *100,* 1450–1453.

129. Sajab, M. S.; Chia, C. H.; Zakaria, S.; Jani, S. M.; Ayob, M. K.; Chee, K. L.; Khiew, P. S.; Chiu, W. S. Citric Acid Modified Kenaf Core Fibres for Removal of Methylene Blue from Aqueous Solution. *Bioresour. Technol.* **2011,** *102,* 7237–7243.

130. Drnovšek, T.; Perdih, A. Selective Staining as a Tool for Wood Fiber Characterization. *Dyes Pigments.* **2005,** *67,* 197–206.

131. He, X.; Male, K. B.; Nesterenko, P. N.; Brabazon, D.; Paull, B.; Luong, J. H. T. Adsorption and Desorption of Methylene Blue on Porous Carbon Monoliths and Nanocrystalline Cellulose. *ACS Appl. Mater. Interfaces.* **2013,** *5,* 8796–8804.

132. Batmaz, R.; Mohammed, N.; Zaman, M.; Minhas, G.; Berry, R.; Tam, K. Cellulose Nanocrystals as Promising Adsorbents for the Removal of Cationic Dyes. *Cellulose.* **2014,** *21,* 1655–1665.

133. Chen, W.; Li, Q.; Wang, Y.; Yi, X.; Zeng, J.; Yu, H.; Liu, Y.; Li, J. Comparative Study of Aerogels Obtained from Differently Prepared Nanocellulose Fibers. *Chem. Sus. Chem.* **2014,** *7,* 154–161.

134. Chan, C. H.; Chia, C. H.; Zakaria, S.; Sajab, M. S.; Chin, S. X. Cellulose Nanofibrils: a Rapid Adsorbent for the Removal of Methylene Blue. *RSC Adv.* **2015,** *5,* 18204–18212.

135. Tatsumi, D.; Ishioka, S.; Matsumoto, T. Effect of Fiber Concentration and Axial Ratio on the Rheological Properties of Cellulose Fiber Suspensions. *J. Soc. Rheol.* **2002,** *30,* 27–32.

CHAPTER 8

QUALITATIVE FOURIER TRANSFORM INFRARED SPECTROSCOPIC ANALYSIS OF POLYETHER-BASED POLYMER ELECTROLYTES

SITI ROZANA BT. ABDUL KARIM and CHIN HAN CHAN[*]

Faculty of Applied Sciences, Universiti Teknologi MARA, Shah Alam 40450, Malaysia

[*]*Corresponding author. E-mail: cchan_25@yahoo.com.sg*

CONTENTS

Abstract 198
8.1   Introduction of Solid Polymer Electrolytes (SPEs) 198
8.2   Thermal Properties and Spectroscopic Analysis of the PEO/Polyacrylate Systems 199
8.3   Qualitative Analysis of FTIR 207
8.4   Qualitative Analysis of FTIR (Band Absorbance) 225
8.5   Summary 237
Keywords 237
References 237

ABSTRACT

Fourier transform infrared (FTIR) is a spectroscopic method that is widely used to study the intermolecular interaction between polymers. In this chapter, the qualitative analysis of FTIR for poly(ethylene oxide)/poly(methyl methacrylate) (PEO/PMMA) based polymer electrolytes are discussed in details. This chapter starts with the discussion on thermal studies of PEO, PMMA, and its blend followed by the qualitative analysis of FTIR for the systems. In this part, the details on step-by-step guidelines, sample of work and discussion focused on deconvolution of absorbance bands for PEO/PMMA systems are presented.

8.1 INTRODUCTION OF SOLID POLYMER ELECTROLYTES (SPES)

The demand for electronic devices such as cell phones, laptops, tablets, etc., with high performance is on the rise nowadays. In order to fabricate energy storage that can support the performance of these devices, extensive researches have been carried out academically and industrially with the ultimate goal to produce better and higher performance lithium-ion rechargeable batteries. Polymer electrolyte is a membrane separator between the cathode and the anode of a rechargeable battery. Basically, there are three types of polymer electrolytes, viz., the gel polymer electrolytes (GPEs), liquid polymer electrolytes (LPEs), and solid polymer electrolytes (SPEs).

SPEs have many advantages over the other two polymer electrolytes because they have better mechanical stability, flexibility for certain polymer hosts, and the ease of fabrication into desirable sizes[1] that can be applied not only in lithium-ion rechargeable batteries but also in fuel cells, electro-chromic windows, super-capacitors, and sensors.[2–4] However, in general, SPEs show low ionic conductivity (σ_{DC}) at room temperature that limit their commercial applications. Efforts to improve the ionic conductivity of SPEs point preferably toward two closely related directions, that is, the enhancement of both charge carrier density and mobility. Most of the binary systems of polymer and salt do not exhibit sufficiently high σ_{DC} ($\sim 10^{-4}$ S cm^{-1}) for applications. Hence, not only homopolymers have been used as polymer host, but polymer blends,[5–8] block copolymers,[9,10] polymer composites,[11,12] etc., have also been used as an effort to enhance the conductivity of SPEs.

To date, many types of ion conducting polymers, viz. PEO,[13–16] PMMA,[17–23] poly(vinylidene fluoride) (PVdF),[24–26] PVC,[27] and poly (acrylonitrile) (PAN)[21,28] have been investigated as hosts for polymer electrolytes. Owing to its relatively simple structure that can provide fast ion transport[29] and the high solvating capability of a wide variety of metallic salts, PEO is extensively studied as a model system in polymer blends and polymer electrolyte.[16,30–35]

Conductivity of a polymer electrolyte is determined by the concentration of charge carrier and its mobility. Ion dissociation is markedly influenced by the lattice energy of the salt[36,37] as well as the solvating ability of the polymer matrix. Ion transport is associated with the amorphous phase of a polymer electrolyte.[38] Segmental motion in polymeric chain is another important criteria governing ion conductivity because it affects the mobility of both cations and anions by controlling the local free volume and viscosity of the environment surrounding charge-transporting ions.[39,40] When an inorganic salt is added to PEO, the cations from the salt form polymer–salt complexes with the ether oxygen of PEO, hence, changing the straight or bent lamellae to randomly oriented lamellae.[41]

8.2 THERMAL PROPERTIES AND SPECTROSCOPIC ANALYSIS OF THE PEO/POLYACRYLATE SYSTEMS

Polymer blending is the most convenient and cost-effective way of designing new polymeric materials with properties not attainable by single polymers, tailored for specific applications.[42,43] Polymer blends can be divided into two categories, i.e., miscible and immiscible blends components. PEO is known to exhibit miscibility with many amorphous polymers via hydrogen bonding between its ether oxygen atom and proton donors such as poly(vinyl alcohol) (PVA),[44] oligoesters (OE), polyester resins (PER),[45] poly(vinyl phenol) (PVPh),[46] and poly(methyl vinyl ether-maleic acid) (PMVE-MAc).[47] However, when blended with non-proton donors, miscibility in the binary PEO-based blend system can also be achieved via relatively weak dipole–dipole interactions.[48–51] The weak intermolecular interactions or strong hydrogen bonds in miscible PEO-based blend tend to hinder the crystallization of PEO, resulting in suppressing its crystallinity and lowering its melting temperature.

The thermal behavior and the spherulitic morphology of PEO/Poly (methyl methacrylate) (PMMA) have been widely explored by many authors.[48,52–56] Other than the extensively documented PEO/PMMA and PEO/poly(methacrylate) (PMA) system, no other acrylate polymers like poly(propyl methacrylate) (PPMA), poly(butyl methacrylate) (PBMA), poly(cyclohexyl methacrylate) (PCMA), etc., are recognized for their miscibility with PEO, and until recently, poly(phenyl methacrylate) (PPhMA),[57] poly(benzyl methacrylate) (PBzMA),[58] poly(n-butyl methacrylate) (PnBMA),[51] PMA,[59] poly(iso-butyl methacrylate) (PiBMA),[60] poly(tert-butyl methacrylate) (PtBMA),[60] and poly(4-vinylphenol-co-2-hydroxyethyl methacrylate) (PVPh-HEM)[61] are found to be miscible with PEO. Owing to the vast differences both in the surface structure and T_gs of PEO and PMMA (T_gs for PEO and PMMA are–52 and 105 °C, respectively), PEO/PMMA blend forms a complex system.[35,62] The two polymers are miscible in the liquid state above lower critical solution temperature (LCST) via weak intermolecular interactions but are immiscible in the solid state at room temperature as PEO crystallizes below 65 °C and PMMA is in its glassy state below 105 °C.

Spectroscopic techniques are particularly useful in examining the degree of phase separation and microstructures of polymer blends.[63–66] Li and Hsu[52] analyzed the crystallization behavior of PEO/PMMA blends using Fourier-transform infrared (FTIR) and calorimetric techniques. On examining the FTIR spectra, it was reported that the crystalline band of 1360 and 1340 cm^{-1} doublet had been replaced by the amorphous 1349 cm^{-1} band with increasing PMMA content, indicating a significant reduction in crystallization rate leading to attenuation in the crystallinity of PEO. The microstructure of PEO is changed from the trans-state to the gauche form attributed to the interaction between oxygen atoms and CH_2 groups. Wang et al.,[67] in their studies on crystalline structure of PEO/PMMA blend film, display weak interaction between the ether oxygen of PEO and the polar methoxy group of PMMA in the blend.

The existence of weak dipole–dipole interactions between PEO and PMMA is also suggested by applying vibrational spectroscopy,[50] SAXS, and SANS.[68] Recently, FTIR spectroscopy has been widely employed to substantiate ramifications on the miscibility of binary blends studied by other techniques.[66,69,70] Most of the authors have suggested that the PEO/PMMA blend is miscible both in the melt[71,72] and in the amorphous phase.[73,74] Further evidence on the miscibility, molecular structure, and local dynamics of PEO/*atactic* PMMA (*a*PMMA) blend is demonstrated by ^{13}C and ^{1}H NMR spectroscopy.[59,71,73–76] Schantz,[49] and Straka et al.,[74] who carried out solid-state ^{13}C NMR measurement on melt-mixed blends of PEO/*a*PMMA, concluded that all the blends with PEO weight fraction $\leq 30\%$ were completely homogeneous in a length scale of 20–70 nm with part of the PMMA and PEO chain intimately mixed in the amorphous phase. The local mobility of amorphous PEO decreases in proportion to the PMMA content added as revealed by the relaxation behavior for protons and carbons. Martuscelli et al.[71] investigated the miscibility of PEO/*a*PMMA using ^{13}C NMR and showed that the linewidth of PMMA signal was strongly dependent on the PEO content and that the PEO mobility was greatly reduced leading to a more rigid blend system with ascending PMMA content.

Quantities glass transition temperature (T_g) has always been an important parameter applied to verify the miscibility of a polymer blends pair. When a binary blend shows two T_gs corresponding to the neat parent polymers, immiscibility may be deduced. On the other hand, a miscible binary blend exhibits one single and composition dependent T_g.[77,78] In the case of PEO/PMMA blends, there are different explanations in literatures on the miscibility behavior of this semicrystalline/amorphous blend.[49,79–81]

The T_gs of neat PEO, PMMA, and its blend of different compositions are extracted from the reheating cycle of the differential scanning calorimetry (DSC) thermogram. Figure 8.1 shows the T_gs of neat PEO and PMMA are at -54 and $105\ °C$, respectively. A single and composition dependent T_g is observed for all blend compositions with increasing content of PMMA in the blend indicates that the PEO/PMMA system is miscible in the molten state under the experimental conditions. It is interesting to note in Figure 8.1 that variation of the experimental T_g values as a function of W_{PEO} agrees closely to that evaluated using the Gordon–Taylor equation but shows a large negative deviation from that calculated using the Fox equation. This correlates with the behavior of most PEO/polyacrylate miscible blends.[57,58] The Fox and Gordon–Taylor equations are given as eq 8.1 and eq 8.2.

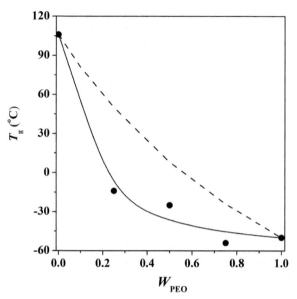

FIGURE 8.1 Variations of glass transition temperature (T_g) of salt-free PEO/PMMA blends as functions of W_{PEO} in solid circle (●). The dashed (---) and solid (——) curves represent the calculated T_gs from Fox and Gordon–Taylor equations, respectively.

Fox Equation:

$$\frac{1}{T_g} = \frac{W_1}{T_{g1}} + \frac{W_2}{T_{g2}}, \tag{8.1}$$

where T_g, T_{g1}, and T_{g2} in Kelvin are the glass transition temperatures of the blend, pure component 1, and pure component 2, while W_1 and W_2 are weight fractions of component 1 and component 2, respectively. However, this method is reliable in ascertaining polymer–polymer miscibility if only the content of the second component is greater than 10%, and the T_gs of the two neat constituents must differ by at least 20 °C.[82,83]

Another equation used to examine the miscibility of a binary blends system is by the Gordon–Taylor Equation as shown in eq 8.3.[84]

$$T_g = \frac{W_1 T_{g1} + k W_2 T_{g2}}{W_1 + k W_2}, \tag{8.2}$$

where W_1 and W_2 are the weight fractions of polymer component and the parameter k is the ratio of the differences of the coefficient of expansion

Qualitative Fourier Transform Infrared Spectroscopic Analysis   203

($\Delta\alpha$) near the glassy and the rubbery states. The constant k can be derived from the experimental data for T_g determination and it is an adjustable parameter. This equation is applied when the miscibility of a polymer blend system is not as simple as described by the Fox equation. The T_gs for complicated system such as PEO/PMMA[49] is in agreement with the T_gs calculated by the Gordon–Taylor equation depicting that this blend system exhibit relatively weak interaction[85].

Owing to the high crystallinity of the semicrystalline PEO and the high T_g of the amorphous PMMA, the PEO/PMMA blend is a complex miscible system which achieve miscibility at temperature range above the melting temperature (T_m) of PEO at 80 °C (in the molten state) up to the LCST of 225 °C as reported in the literature.[86–88] When the PEO/PMMA blend system is cooled down from the molten state to approximately 105 °C, PMMA forms a glassy state, at temperature below 80 °C, liquid–solid phase separation is observed, where PEO crystallizes with PMMA in the glassy state. However, a homogeneous mixture of PEO and PMMA exists in the amorphous phase of the cooled-down system which forms the ion percolation pathway for the transportation of the Li^+ when salt is added. In general, the PEO/PMMA blends are known to be miscible with relatively weak dipole–dipole interactions both in the molten state and in the amorphous phase of the solid PEO/PMMA blends at room temperature.

The T_g values for neat PEO is observed to increase gradually with ascending $LiClO_4$ up to salt concentration $(Y_S) = 0.10$ whereas the T_g values of PMMA increase at low $LiClO_4$ content, then plateau at $Y_S = 0.02$ as shown in Figure 8.2. Addition of $LiClO_4$ does not affect the miscibility of the amorphous phase of PEO/PMMA 75/25 and 25/75 blends at room temperature as shown by the presence of one T_g throughout the salt concentration up to $Y_S = 0.12$. The T_g values of PEO/PMMA 75/25 increase monotonically with ascending salt content until $Y_S = 0.05$, then remain constant up to $Y_S = 0.10$ before decreasing at higher salt concentration. Besides, The T_g values for the PEO/PMMA 75/25 blend almost overlap those of the PEO/$LiClO_4$ system at $Y_S < 0.05$ but record a higher value at salt concentration ≥ 0.05 indicates that more Li^+ ions are strongly coordinated to the ether oxygen of PEO in the PEO/PMMA/$LiClO_4$ systems as compared to the PEO–salt system at $Y_S = 0.05$–0.10. On the whole, it can be deduced from Figure 8.2 that Li^+ ions interact strongly with PEO but so with PMMA. Formation of the polymer–Li^+ ion complexes decreases the flexibility of the polymer chains resulting in an increase in the T_g values of the blend.

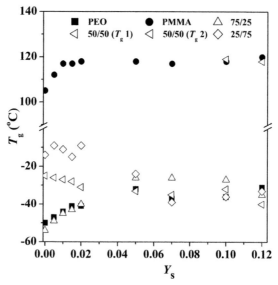

FIGURE 8.2 Glass transition temperature (T_g) of PEO/PMMA systems at different salt concentrations.

Interestingly, two T_gs with close proximity to the neat PEO and PMMA are observed for the PEO/PMMA 50/50 system at $Y_S \geq 0.10$. This observation may be due to the addition of a large amount of salt ($Y_S \geq 0.10$) to the PEO/PMMA 50/50 system which triggers liquid–liquid phase separation in the molten state of the blend. Furthermore, when the salt-added PEO/PMMA 50/50 system is cooling down to room temperature, additional liquid–solid phase separation takes place which may cause an immiscible mixture of PEO, PMMA, and salt in the amorphous phase of the solid thin film.

The melting temperature and the melting enthalpy (ΔH_m) for each samples are obtained from the maximum peak and the area under the peak of the reheating cycle, respectively. Figure 8.3 presents the depression of T_m in PEO/PMMA and PEO/PMMA/LiClO$_4$ systems with increasing content of PMMA, respectively. From the thermal curve, only one melting peak is observed for all the samples. Depression of T_m of PEO with increasing PMMA content in both the salt-free and salt-added PEO/PMMA systems corroborates with the T_g result which points toward miscibility between PEO and PMMA. In Figure 8.3, it is observed that addition of salt causes further depression of T_m of PEO at each composition of the PEO/PMMA blend suggesting better ion-interaction of the salt with PEO as compared to PMMA.[5,49]

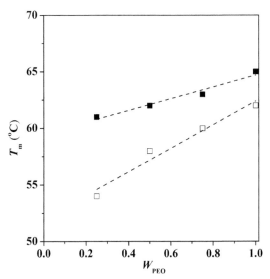

FIGURE 8.3 Melting temperature (T_m) of PEO in salt-free (■) and salt-added (□) PEO/PMMA systems. The dashed curve marked the depression of T_m in the systems.

The selected values of T_m and ΔH_m for the salt-free and the salt-added PEO and PEO/PMMA systems presented in Table 8.1 depict that both the T_m and ΔH_m values of PEO in the selected blends decrease with ascending PMMA content as well as salt concentration. The crystallinity (X^*) of PEO in neat PEO and selected PEO/PMMA blend compositions with and without LiClO$_4$ was determined using eqs 8.3 and 8.4 for the salt-free and salt-added blends, respectively.

TABLE 8.1 T_m and ΔH_m for Selected PEO/PMMA/LiClO$_4$ Systems.

Y_S	PEO/PMMA							
	PEO		75/25		50/50		25/75	
	T_m (°C)	ΔH_m (J·g^{-1})	T_m (°C)	ΔH_m (J·g^{-1})	T_m (°C)	ΔH_m (J·g^{-1})	T_m (°C)	ΔH_m (J·g^{-1})
0	64.7	130.9	62.9	82.4	62.1	52.7	61.5	14.5
0.05	61.7	111.6	60.3	72.2	58.4	36.1	54.2	12.5
0.10	57.9	80.5	56.8	60.2	49.0	17.0	–	–
0.12	57.6	71.7	54.2	48.5	47.3	14.7	–	–

$\Delta H_{ref}^{\circ} = 188.3$ J g^{-1} denotes 100% crystallinity of PEO.

$$X^* = \left(\frac{\Delta H_m}{\Delta H_{ref}^{\circ}}\right) \text{ for salt-free blends} \qquad (8.3)$$

$$X^* = \left(\frac{\Delta H_m}{(1-Y_S)\Delta H_{ref}^{\circ}}\right) \text{ for salt-added blends.} \qquad (8.4)$$

The plots of X^* of the PEO/PMMA and PEO/PMMA/LiClO$_4$ ($Y_S = 0.05$) systems versus W_{PEO} as demonstrated in Figure 8.4 depict a negative deviation from the constancy of the crystallinity of PEO in the salt-free PEO/PMMA which reflects progressive reduction in the crystallinity of PEO with ascending PMMA content. Again, the crystallinity result reaffirms the miscibility of PEO and PMMA in the amorphous phase of the solid thin film at room temperature. Meanwhile, addition of a small amount of salt ($Y_S = 0.05$) causes only a slight dip in the X^* of neat PEO.[89]

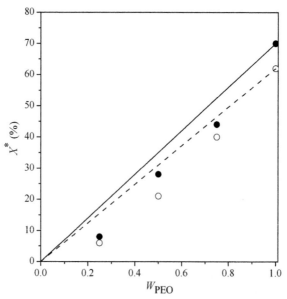

FIGURE 8.4 Plots of Crystallinity (X^*) of PEO in the salt-free (●) and salt-added ($Y_S = 0.05$) (○)PEO/PMMA systems versus W_{PEO}. The solid and dashed curves mark the constancy of the crystallinity of the PEO/PMMA and PEO/PMMA/LiClO$_4$ blends, respectively.

Qualitative Fourier Transform Infrared Spectroscopic Analysis   207

Similar reduction in the X^* of PEO concomitant to that of the salt-free PEO/PMMA systems is also observed for the blends doped with salt concentration $Y_S = 0.05$. In addition, the X^* values of the PEO/PMMA/LiClO$_4$ system as shown in Figure 8.4 are very close to those of the salt-free PEO/PMMA system, which indicates that the X^* of PEO in the blends is not influenced by the presence of the salt. Therefore, it can be concluded that most of the salt added dissolves in the amorphous phase of PEO rather than PMMA in the PEO/PMMA system which is in good agreement with the T_g results discussed earlier.

Although there were lots of studies on the thermal and electrical properties of SPEs, the effective role of the lithium salt in the affecting miscibility of polymer blends as well as promoting ion transport and also enhancing the conductivity is not yet well understood. Many models especially on the interaction between the salt and the polymer chains have been proposed to explain the role played by the salt in single polymer–salt electrolyte. But, not many researchers have seriously looked into the relationship between the conductivity mechanism as well as the interfacial stability and the distribution of the lithium salt SPEs consisting of polymer blend electrolyte.

Therefore, in order to investigate and understand the mechanism as well as distribution of LiClO$_4$ in the two components of the blend and the role it played on influencing the properties of PEO/PMMA blend, FTIR spectroscopy technique has been chosen as it enables to detect any changes in the molecular level as it is highly sensitive and the spectrum from the FTIR can be used to analyze both quality and quantity measurements.

8.3 QUALITATIVE ANALYSIS OF FTIR

8.3.1 INTRODUCTION OF FTIR ANALYSIS

FTIR spectroscopy is a powerful tool to analyze the ion-molecular interaction between polymers. Basically, the technique of infrared spectroscopy is based on the vibrations of atoms in molecules when infrared radiation passed through a sample. The evolution of infrared improved dramatically in terms of the quality and time required to obtain data when established mathematical process of Fourier-transform was introduced.

There are different types of infrared sampling technique intended to be used for different types of samples and one of these techniques is attenuated total reflectance (ATR). This technique is used to analyze solids, liquids, semisolids, and thin films.[90] ATR needs very minimal sample preparation and it is a non-destructive sampling technique. In ATR, an accessory is mounted in the sample compartment of FTIR. At the center of the ATR base assembly is a crystal (e.g., diamond, germanium, and zinc selenide (ZnSe)) which is used to reflect the IR beam from the source to the sample and then to the detector. Figure 8.5 shows the schematic diagram of an ATR base assembly.

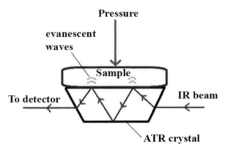

FIGURE 8.5 A schematic diagram of an attenuated total reflectance base assembly.

Polymer–polymer, polymer–salt, polymer–filler and polymer–salt–filler interactions for PEO, PMMA, and their blends with and without LiClO$_4$ at room temperature were studied using FTIR (Perkin Elmer Spectrum One, USA) coupled with an ATR base assembly with diamond crystal window. The sample was scanned at frequency range of 600–4000 cm^{-1} at a resolution of 2 cm^{-1} with 16 scans per spectrum. The assignment of absorption band for LiClO$_4$, PEO, and PMMA are listed in Tables 8.2–8.4, respectively.

TABLE 8.2 Assignment of Absorption Band of FTIR for LiClO$_4$.

Assignment	Wavenumber (cm^{-1})	Reference
Pure LiClO$_4$	1630, 1368	[7]
	1610, 1300	[91]
	1610, 1070	[92]
$v_{as}(ClO_4^-)$	1089	[7]
ClO$_4^-$ anion	930	[81]
doublet, spectroscopically "ion pair" and "free" ClO$_4$, respectively	633, 616	[7]

Qualitative Fourier Transform Infrared Spectroscopic Analysis

TABLE 8.3 Assignment of Absorption Bands of FTIR for PEO.

Assignment	Wavenumber (cm⁻¹)	Reference
$v_{as}(CH_2)$, $v_s(CH_2)$	2946, 2881	[7]
CH streching	2900	[91]
	2806	[7]
v_{as} of PEO	1950	[91]
$\delta(CH_2)$	1485	[91]
$\delta_{as}(CH_2)$, $\delta_s(CH_2)$	1466, 1456	[7]
α_{as} of PEO	1450	[91]
A doublet, $\omega(CH_2)$	1360, 1341	[7]
$\tau_{as}(CH_2)$, $\tau_s(CH_2)$	1279, 1240	[7]
$v_s(C-O-C)$	1145, 1093, 1060	[7]
$v(C-C) + \rho(CH_2)$	1060	[7]
$\tau(CH_2)$	991	[91]
$\rho_{as}(CH_2) + v(CH_2)$, $\rho_s(CH_2) + v(C-C)$	960, 945	[7]
$\omega(CH_2)$	842	[7]
$\rho(CH_2) + \delta(C-O-C)$	841	[7]
$\rho(CH_2)$	749	[91]

TABLE 8.4 Assignment of Absorption Bands of FTIR for PMMA.

Assignment	Wavenumber (cm⁻¹)	Reference
$v_{as}(CH_3)$	3010	[93]
$v_{as}(CH_2)$	2954	[94]
	2951	[93]
$v(CH)$	2927	[95]
$v_s(C=O)$	1725	[96,97]
$\alpha_{as}(CH_3)$	1485	[98,99]
	1452	[93]
$\rho(CH_2)$	1483	[93,94]
$\delta(CH_2)$	1450	[99]
$v(CH_3)$	1449	[70]
$\delta(O-CH_3)$	1430	[98]
	1389	[93]
$v(C-CH_3)$	1368	[99]
$v(C-O)$	1280	[93,94]

TABLE 8.4 *(Continued)*

Assignment	Wavenumber (cm^{-1})	Reference
$\tau(CH_2)$	1279	[99]
$v(C-C-O)$	1250	[100]
$v_s(C-O-C)$, $v_{as}(C-O-C)$	1242, 1188	[97]
$v(C-O)$ of the $O-CH_3$ group	1145	
$v_{as}(C-O-C)$	1240, 1145	[96]
$v(C-O)$ of the $O-CH_3$ group	1190	
$\alpha(C-O-C)$	1180	[100]
$\tau(CH_2)$	1173	[93,94]
$v(C-C)$ in $C-C-O$	1171	[101]

8.3.2 GENERAL RULE OF THUMB TO DO QUALITATIVE FTIR ANALYSIS

After data acquisition, the next step is to analyze the spectrum. There are different approaches to analyze the FTIR result. Qualitative analysis approach is relatively simple to deduce useful information for this study. The objective to do the qualitative analysis is to identify functional groups and characterizing the intermolecular interaction that exist in the sample based on the spectral band. It is crucial for one to take care of the data-processing steps in order to obtain more precise data interpretation is subjected.

There are many software packages that can be used to process the FTIR data such as SPECTRUM, OMNIC, etc., depending on the brand of FTIR spectrometer used. In any FTIR software, the basic functions that are usually used to improve the quality of a spectrum are baseline correction, smooth, labeling, stacking spectra, and exporting data to Microsoft Excel. After exporting data to the Excel file, one can proceed to further analyze the spectrum in Origin™ or any graphical program that can assist in the further analysis of the spectra.

From the FTIR spectra, one has to select a few regions that are suitable for further data processing. However, it is always challenging to decide which IR regions to be scrutinized further. It is important to examine the whole mid-IR region at the beginning. For instance, consider the case of salt-free and salt-added PEO/PMMA blend. It is important to analyze each of the functional groups for both polymers in the absence and presence of $LiClO_4$ in order to understand the interaction between the polymer and the salt. In this case, one needs to start the analysis from the highest (e.g., $-OH$ at ~3500 cm^{-1}) to the lowest wavenumber (e.g., ClO_4^- at ~650 cm^{-1}). From here, the unchanged bands (e.g., band shape and band intensity) that are not affected by the addition of $LiClO_4$ can be excluded from further discussion.

8.3.3 INTRODUCTION OF FTIR SPECTRA OF PEO, PMMA, AND LICLO$_4$

FTIR is a spectroscopic method which is widely used to study the intermolecular attractions between components of a polymer blend[7,74] and also the ion–dipole interaction between salt and polymer.[14,63,81] FTIR is a useful tool as it provides information especially on compatibility of blends and the salt–polymer interaction which corroborates other studies such as DSC, POM, and also XRD. The absorption bands of a system obtained from this technique give insight on the ion–dipole or dipole–dipole for the system studied. Change in intensity and shifting of the absorbance bands for a FTIR spectrum have always been the main focus of the data analysis in order to understand the mechanism of interaction inside a polymer chain.

The PEO, PMMA, and $LiClO_4$ used in this study are well known as many studies have been done on the molecular interaction of salt-free and salt-added PEO/PMMA systems. We first show the absorbance bands for pure components for the confirmation of the component. Subsequently, an extensive investigation on FTIR and the relationship with other analyses are discussed in order to verify precisely the molecular interaction and also the effect of ion–dipole interaction to the miscibility of PEO/PMMA systems as discussed in the earlier section.

Study on miscibility of salt-free and salt-added PEO/PMMA systems using DSC found that the salt-free PEO/PMMA systems are miscible as one T_g is observed throughout the composition. For the salt-added PEO/PMMA systems, the PEO/PMMA 75/25 and PEO/PMMA 25/75 systems show one T_g for all salt concentrations but PEO/PMMA 50/50 system was found to phase separate when $Y_S > 0.07$. From the thermodynamic point of view, addition of high concentration of salt that triggers solid–liquid phase separation is due to the very weak interaction of PEO/PMMA. In order to support this deduction, FTIR is a suitable technique to be employed for better understanding on both dipole–dipole and ion–dipole interactions between PEO, PMMA, and LiClO$_4$.

The FTIR spectra and characteristic bands of PEO are presented in Figure 8.6 and Table 8.5, respectively. As reference, the symbol for vibrational mode used in the table and discussion are given as; v = stretch, v_s = symmetrical stretch, v_{as} = assymetrical stretch, α = bend, δ = scissor, ω = wag, ρ = rock, τ = twist. Meanwhile, the intensities are given as vs = very strong, s = strong, m = medium, and w = weak. From Figure 8.6, the $\omega(CH_2)$ mode of PEO which appears as a sharp doublet at 1359 and 1341 cm^{-1} denotes the crystalline phase of PEO. The doublet showing strong intensities indicates that PEO is highly crystalline at 30°C. On top of that, the triplet of v_s (C–O–C) also exhibiting strong intensities denotes that there is no intermolecular interaction occurs in this group.

TABLE 8.5 Characteristic Bands of PEO.

No.	Wavenumber (cm^{-1})	Assignment
1	841 s	$\rho(CH_2) + \delta(C-O-C)$
2	961 s, 947 s	$\rho_{as}(CH_2) + v(CH_2), \rho_s(CH_2) + v(C-C)$
3	1065 s, 1096 vs, 1144 s	(triplet, with maximum at 1096 cm^{-1}) $v_s(C-O-C)$
4	1241 s, 1279 s	$\tau_s(CH_2), \tau_{as}(CH_2)$
5	1342 s, 1360 s	doublet, $\omega(CH_2)$, represent the crystalline phase
6	1455 m, 1467 s	$\delta_s(CH_2), \delta_{as}(CH_2)$
7	2882 m	$v(CH)$

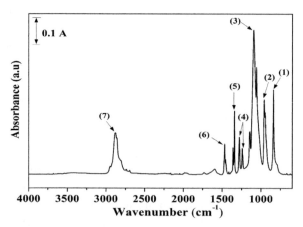

FIGURE 8.6 FTIR spectrum of PEO.

Figure 8.7 and Table 8.6 represent the FTIR spectra and characteristic bands for PMMA, respectively. PMMA has carbonyl group which is easily indicated from the spectra at 1723 cm^{-1}. This band is sharp and has very strong intensities. Other than that, the C–O–C band of PMMA which exhibits a maximum at 1143 cm^{-1} also demonstrates sharp and high intensity points toward the purity of PMMA. The observation from the neat PEO and PMMA spectra confirms the purity of polymer used in this study.

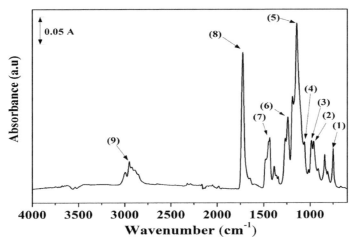

FIGURE 8.7   FTIR spectrum of PMMA.

TABLE 8.6   Characteristic Bands of PMMA.

No.	Wavenumber (cm^{-1})	Assignment
1	749 s	$\rho(CH_2)$
2	964 s	$\omega(CH_2)$
3	988 s	$v_s(C-O)$ in the C–O–C
4	1061 m	$v(O-CH_2)$
5	1143 vs, 1190 s	$v(C-O)$ of $O-CH_3$
6	1240 s	$v_s(C-O-C)$
7	1435 m	$\delta(O-CH_3)$
8	1723 vs	$v(C=O)$
9	2950 m	$v_{as}(CH_2)$

The LiClO$_4$ has very simple FTIR spectra as shown in Figure 8.8. The vibrational band characteristic for LiClO$_4$ is shown in Table 8.7, confirming the purity of the salt used. As depicted in Figure 8.8, the 616 cm^{-1} band, which attributes to the less associated ClO$_4^-$, exhibits high intensity indicates that the LiClO$_4$ is highly soluble in the polymer used in this study. The dissociation of the salt is presented schematically as Li$^+\cdots$ClO$_4^-$.

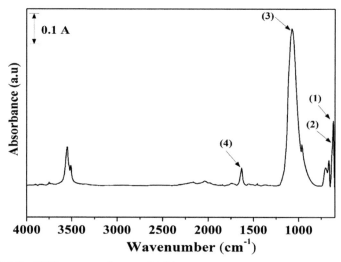

FIGURE 8.8 FTIR spectrum for pure LiClO$_4$.

TABLE 8.7 Characteristic Bands for LiClO$_4$.

No.	Wavenumber (cm^{-1})	Assignment
1	617 s	v(ClO$_4^-$); less associated ClO$_4^-$
2	630 w	v(ClO$_4^-$); more associated ClO$_4^-$
3	1077 vs.	v_{as}(ClO$_4^-$)
4	1632 m	pure LiClO$_4$

8.3.4 ANALYSIS ON FTIR SPECTRA OF PEO/LICLO$_4$ SYSTEM

When LiClO$_4$ is added to the PEO/PMMA systems, the band at 616 cm^{-1} is shifted to 623 cm^{-1} and the intensity of band at 630 cm^{-1}, which represents the more associated ClO$_4^-$, reduces. The variation of this phenomenon is depicted in Figure 8.9 which represents salt-added PEO system at the wavenumber range of 640–600 cm^{-1}. Other than that, the addition of salt to PEO creates changes on the selected vibration mode especially in v(C–O–C) region as presented in Figure 8.10. The triplet v(C–O–C) with maximum at 1096 cm^{-1} is shifting to lower wavenumber at 1079 cm^{-1} when $Y_S \geq 0.10$. The two shoulders' vibration at 1144 and 1060 cm^{-1} become weaker when salt concentration is increased. The shifting reveals that there is some interaction between ether oxygen and salt. The high electronegative ether oxygen (–O–) of PEO attracts the Li$^+$ ion from the salt to form interaction. The changes are obvious for the band at 1096 cm^{-1} when salt is added as compared to the two shoulders, explaining that the salt prefers to form interaction with oxygen in the amorphous part of PEO. This observation is supported with the ω(CH$_2$) at region of 1370–1320 cm^{-1} in Figure 8.11. The doublet (1360 and 1342 cm^{-1}) shows no shifting in wavenumbers throughout the salt concentration but shows reduced intensity and broadening when $Y_S \geq 0.12$ depicting that the crystal structure of PEO is affected at high amount of salt due to the weak polarization of CH$_2$ with the salt. The possible coordination of the salt with the oxygen is shown in Figure 8.12.

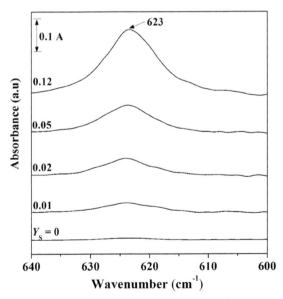

FIGURE 8.9 FTIR spectra at ClO$_4^-$ region for salt-added PEO system.

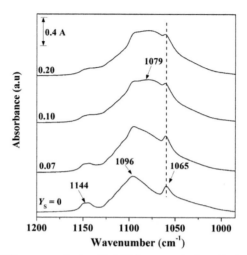

FIGURE 8.10 FTIR spectra of salt-added PEO system in the ν(C–O–C) region.

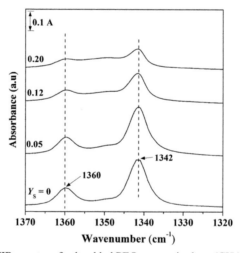

FIGURE 8.11 FTIR spectra of salt-added PEO system in the ω(CH$_2$) region.

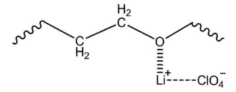

FIGURE 8.12 Schematic diagram of possible coordination of LiClO4 in salt-added PEO system.

8.3.5 ANALYSIS ON FTIR SPECTRA OF PMMA/LICLO$_4$ SYSTEM

Unlike PEO, PMMA does not have highly electronegative site, thus lowering the chances of interaction between the polymer and the salt. However, it is still important to investigate the possible coordination occurred along the PMMA chain in order to have a better understanding on the ionic conductivity which is crucial in the discussion on the PEO/PMMA systems. Addition of salt to PMMA shows some changes in selected vibration bands of PMMA. The $v(C=O)$ at 1723 cm^{-1} is the most prominent band in PMMA and observation on this region has found that the band at 1723 cm^{-1} does not shift in wavenumber with addition of salt as shown in Figure 8.13. However, the intensities of the band reduced and a new band appeared very weakly at 1649 cm^{-1} when $Y_S \geq 0.10$. This new band may attribute to the interaction between the carbonyl oxygen and the salt. Analysis on $v(C–O–C)$ of PMMA with different concentrations of salt shows no change in intensities, but slight shifting to a higher wavenumber is recorded when salt is added (see Fig. 8.14). This phenomenon may be indicating that there is interaction between the oxygen of C–O–C and the salt. Figure 8.15 shows the possible coordination of PMMA and salt.

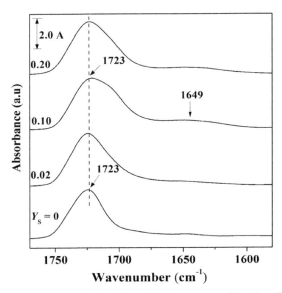

FIGURE 8.13 FTIR spectra of salt-added PMMA system at $v(C=O)$ region.

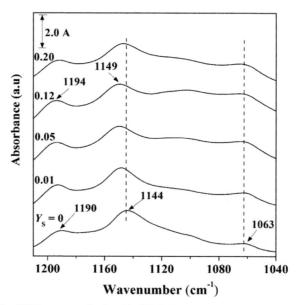

FIGURE 8.14 FTIR spectra of salt-added PMMA system at ν(C–O–C) region.

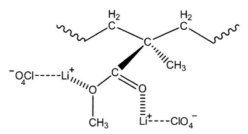

FIGURE 8.15 Schematic diagram of possible coordination of salt-added PMMA system.

8.3.6 ANALYSIS ON FTIR SPECTRA OF SALT-FREE AND SALT-ADDED PEO/PMMA BLEND

In the case of blending of PEO with PMMA, there is no shifting in the position of the ω(CH$_2$) mode as shown in Figure 8.16. However, a slight broadening of the doublet occurs when the PMMA content reaches 50 wt% (W_{PEO} = 0.5). The ν(C–O–C) mode occurs as a triplet centered at 1096 cm^{-1} which represents the amorphous region of PEO while the two shoulders at 1144 and 1065 cm^{-1} characterize the crystalline phase of PEO. When W_{PEO}

in the blend = 0.5, the center band 1096 cm^{-1} starts to broaden without shifting in position and the shoulder at 1060 cm^{-1} diminishes in intensity whereas the shoulder at 1143 cm^{-1} remains unchanged. The strong transmittance intensity of the band at 1143 cm^{-1} could be due to overlapping with the center band of the v(C–O–C) mode of PMMA. On the other hand, no obvious changes are observed at the absorbance band at 1723 cm^{-1} corresponding to the v(C=O) of PMMA.

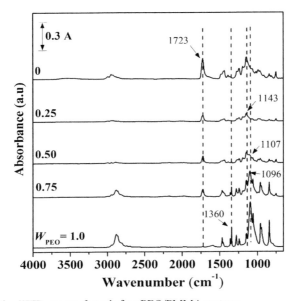

FIGURE 8.16 FTIR spectra for salt-free PEO/PMMA systems.

As one can see from the FTIR spectra of salt-added PEO and PMMA, both polymers have the "preferred" site for the salt to form interaction with the electron donor group in the polymer. Therefore, for the salt-added PEO/PMMA systems, investigation was done to study the salt dispersion in different phases and also competition that might be presence in the systems. There is no shifting observed for bands at 1189 and 1144 cm^{-1}, which represent the C–O–C of PMMA, confirms that there is lacking of interaction occurred between the Li$^+$ ion and PMMA in the presence of major component of PEO. Thus, the salt is more likely to form coordination with ether oxygen of PEO as it is more electronegative as compare to PMMA.

The effect of major component in a blend on the ion–dipole interaction can be seen in selected FTIR region, which based on the earlier analysis focused on the v(C–O–C) and v(C=O) regions. Figure 8.17(a) and (b) shows the v(C–O–C) region of PEO/PMMA 75/25 and PEO/PMMA 25/75 systems, respectively. Figure 8.17(a) presents the v(C–O–C) of PEO/PMMA 75/25 system. For the blend, the vibration at 1098 cm^{-1}, which attributes to the amorphous PEO, is seen very weakly even though the amount of PEO in the blend is quite high. This might be explained by the existence of dipole–dipole interaction between PEO and PMMA. However, this explanation needs further elucidation. The addition of small amount of salt (Y_S = 0.02) causes the band at 1098 cm^{-1} to shift to the low wavenumber which indicates that the Li$^+$ ion the v(C–O–C) band for PEO/PMMA 25/75 system with addition of salt. The two bands of C–O–C stretching of PMMA at 1196 and 1145 cm^{-1} are clearly seen up to Y_S = 0.12. The center band of amorphous C–O–C of PEO is not seen in this blend as PEO is the minor component. The band at 1060 cm^{-1} which attributed to the crystalline C–O–C of PEO is very weak and addition of salt cause the band to diminish and form broad band when the salt added is at Y_S = 0.20.

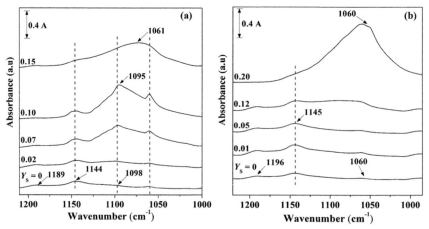

FIGURE 8.17 FTIR spectra at v(C–O–C) region for PEO/PMMA systems (a) 75/25 and (b) 25/75 with different concentration of salt.

Figure 8.18(a) shows the v(C=O) of PEO/PMMA 75/25 system with addition of LiClO$_4$. From the figure, the carbonyl band at 1724 cm^{-1} shifts to the higher wavenumber region and the intensities decrease with increase in LiClO$_4$ concentration. Along with the disappearance of 1724 cm^{-1} band, a new band appears at 1646 cm^{-1} when the salt added at $Y_S = 0.10$. The carbonyl band completely diminishes when the salt added is $Y_S = 0.30$ and the new band shifts to lower wavenumber region. The appearance of the new band is maybe due to the interaction of oxygen from the C=O with the Li$^+$ ion. However, the appearance of new band is only observed when the amount of salt added is high, which may attribute that Li$^+$ is more preferable to form interaction with PEO as compared to PMMA. Figure 8.18(b) shows the v(C=O) of PEO/PMMA 25/75 system with addition of salt. From the figure, the addition of salt affects the carbonyl stretching when the addition of salt is at $Y_S \geq 0.10$. A small hump is observed, however, the intensities are not as strong as PEO/PMMA 50/50 system. This might be explained by the interaction strength between the PEO, PMMA, and LiClO$_4$. In the blend with high PEO or PMMA, the two polymers are mixed very well in the amorphous phase as can be seen from the disappearance of 1098 cm^{-1} which belong to the amorphous PEO. Small amount of PEO and highly compatible interaction of the two polymers cause the Li$^+$ ion along the miscible polymer chains to move toward the carbonyl oxygen. Figure 8.19 and Figure 8.20 below show the schematic diagram for PEO/PMMA 75/25 and PEO/PMMA 25/75 systems with addition of salt, respectively.

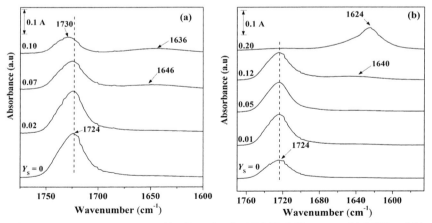

FIGURE 8.18 FTIR spectra at v(C=O) region for PEO/PMMA systems (a) 75/25 and (b) 25/75 with different concentration of salt.

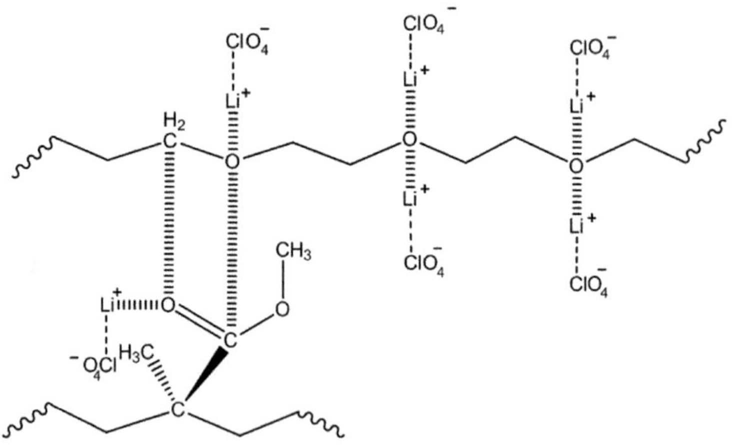

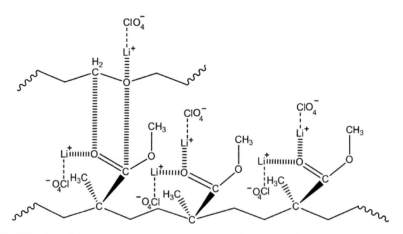

FIGURE 8.19 Schematic diagram on the complexation of LiClO$_4$ to the oxygen of PEO/PMMA 75/25 system.

FIGURE 8.20 Schematic diagram on the complexation of LiClO$_4$ to the oxygen of PEO/PMMA 25/75 system.

Figure 8.21 shows the v(C–O–C) band of PEO/PMMA 50/50 system with addition of salt. From the figure, the band at 1062 cm^{-1} indicates the crystalline phase of PEO disappears when the amount of salt added is at $Y_S \geq 0.07$. However, no change is observed with the C–O–C band of PMMA at 1190 and 1142 cm^{-1} throughout the salt concentration indicates that the Li$^+$ does not form interaction with the oxygen from PMMA. On the other hand, the band at 1106 cm^{-1} which represent the amorphous part of C–O–C PEO is clearly seen even at neat and the appearance of this band is more obvious in this composition as compared to other blend compositions (PEO/PMMA 75/25 and 25/75). This might be explained by the dipole–dipole interaction between PEO and PMMA. Addition of salt to this weak dipole–dipole interaction has led to the coordination of Li$^+$ ion with the ether oxygen of PEO, which is much stronger. Thus it may cause phase separation of PEO and PMMA when the salt concentration is higher. The broaden and shifting of 1106 cm^{-1} band at $Y_S \geq 0.07$ indicates that the ion-polymer interaction is first occur in PEO. The exhaustion site of C–O–C PEO causes some of the Li$^+$ goes to the PMMA, specifically at carbonyl site. Figure 8.22 presents the $v(C=O)$ of PMMA in the blend with addition of salt. The sharp band at 1723 cm^{-1} does not shift in vibration wavenumber, but it decreases with increases in salt content. The new band at 1646 cm^{-1} which is the complex of Li$^+$ and carbonyl oxygen starts to appear at $Y_S = 0.07$ and the band becomes sharper as the salt concentration increased up to $Y_S = 0.30$. The coordination of Li$^+$ ion with PMMA at the carbonyl oxygen and also with PEO at the ether oxygen points toward that in this blend, the salt prefers to form interaction with both polymers. The schematic diagrams on the coordination of salt to PEO and PMMA in low and high salt concentration are presented in Figures 8.23 and 8.24, respectively.

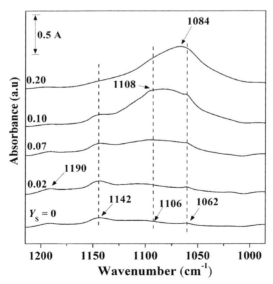

FIGURE 8.21 FTIR spectra at $v(C–O–C)$ region for PEO/PMMA 50/50 system with different concentration of salt.

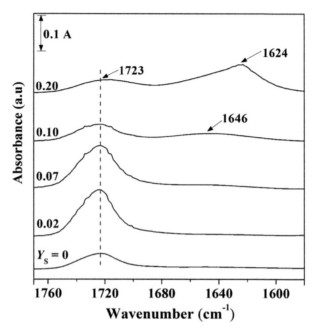

FIGURE 8.22 FTIR spectra at $\nu(C=O)$ region for PEO/PMMA 50/50 system with different concentration of salt.

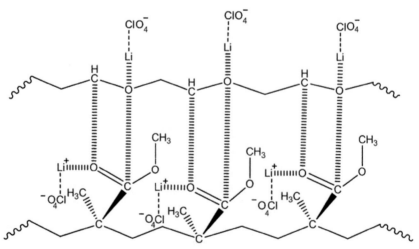

FIGURE 8.23 Schematic diagram on the complexation of LiClO$_4$ at low concentration to the oxygen of PEO/PMMA 50/50 system.

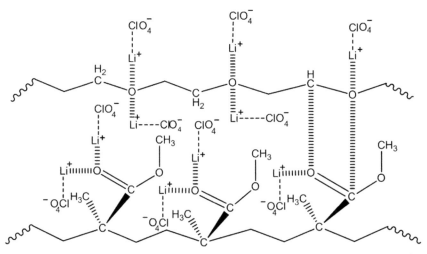

FIGURE 8.24 Schematic diagram on the complexation of LiClO$_4$ at high concentration to the oxygen of PEO/PMMA 50/50 system.

8.4 QUALITATIVE ANALYSIS OF FTIR (BAND ABSORBANCE)

Other than qualitative analysis through identification of functional groups, analyzing variation of the absorbance of a band is a practical way to study the changes in FTIR spectra in more detail manner, for example, to elucidate the areas of different bands for not well resolve absorption bands by deconvolution. Even though deconvolution approach related to in some cases is quantitative analysis, however, this approach is useful to be used qualitatively to estimate degree of molecular interaction of selected functional groups for SPEs. The practical approach and some tips will be discussed in the subsequent sections.

8.4.1 INTRODUCTION OF DECONVOLUTION OF FTIR ABSORBANCE BANDS

Deconvolution or curve fitting is a mathematical style to resolve two or more unresolved combined bands on a section of a spectrum. The purposes of this technique are (1) to enhance the resolution of a spectrum[104] and (2) to give a quantitative/qualitative analysis by fitting the curves rather than obtaining the fundamental fits. The process of deconvolution usually retains band positions (but this may not be applicable in all cases); however, altered the band shapes and band areas of a spectrum. The example of perchlorate ion (ClO_4^-) bands before and after the deconvolution process is shown in Figure 8.25.

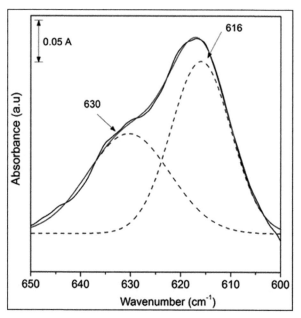

FIGURE 8.25: The ClO_4^- spectrum for PEO at Y_s = 0.02. The solid black curve is the original curve for the unresolved ClO_4^- region, the dash curve shows band after deconvolution, meanwhile the solid red represents the sum of peaks after deconvolution.

8.4.2 STEP-BY-STEP GUIDELINES ON THE DECONVOLUTION OF FTIR ABSORBANCE BANDS AND SAMPLE OF WORK ON THE DECONVOLUTION FOR PEO/PMMA/LICLO$_4$ SYSTEM

Before deciding to perform deconvolution, one is advisable to know and have basic understand on their FTIR spectra. Deconvolution is suitable for spectrum that has two or more mixtures of materials and selected functional groups in the spectrum that show changes. Besides, one has to estimate the number of absorption bands and positions of each band of the overlapped bands before the execution of deconvolution.

The first step before performing deconvolution is to choose a region or a section of a spectrum containing bands to be resolved,[80] in this case, the ClO_4^- region is selected. Different mathematical or graphical software packages are available to assist in performing deconvolution. In this study, Origin Pro is used to deconvolute the selected bands in all samples. This software has suitable feature under "multiple band fit" section that allows one to choose the band type and number of bands to deconvolute. There are different line shape functions which are Gaussian, Lorentzian, and Voigt functions. The most typical function used to deconvolute FTIR band is Gaussian, however, one can choose either one of the band type depending on the suitability of the band. The detail on how to do deconvolution is discussed in the next paragraph. After completing the process, the data (area, center, width, and height) from the report are extracted for further analysis. From these data, qualitative study of a functional group (e.g., O–H, C=O, and ClO_4^-) which can be determined by assuming that only "less associated" and "more associated" forms are distinguishable in the band using the following eqs 8.5 and 8.6:

To determine fraction of less associated functional groups: [103]

$$ f_f = \frac{A_f}{A_f + A_b / a} \tag{8.5} $$

To determine fraction of more associated functional groups:[105]

$$ f_b = \frac{A_b / a}{A_b / a + A_f} \tag{8.6} $$

Where f_f and f_b are fraction of less and more associated functional groups, respectively, meanwhile A_f and A_b are areas (absorbances) under the bands representing less and more associated functional groups, respectively. The constant "a" is the ratio of the molar absorption coefficient which represent in eq 8.7 below. Quantity f_f and f_b give some information on the fraction of each composition in the spectrum.

$$ a = \frac{\text{wavenumber of more associated functional groups}}{\text{wavenumber of less associated functional groups}} \tag{8.7} $$

To give more understanding, below shows the sample of calculation of ClO_4^- region for PEO at $Y_s = 0.02$:

$$a = \frac{630}{616}$$
$$= 1.02$$

$$f_f = \frac{96.0}{96.0 + 4.0 / 1.02}$$
$$= 0.961$$
$$= \underline{\underline{96.1\ \%}}$$

$$f_b = (100 - 96.1)\%$$
$$= \underline{\underline{3.9\ \%}}$$

The general procedure of performing deconvolution starts with any commercial FTIR software, for example SPECTRUM software. First, open new file by clicking File, then Open file (*.sp for spectra). From here, one can select more than one file to open the spectra. Usually, the spectrum is in transmittance mode. In order to perform deconvolution, the spectrum shall be changed to absorbance mode. To change it, go to Process, then Absorbance. In order to get the data for the selected region or band to deconvolute, each spectrum will be exported to a new Excel file. The file will be in Comma Separated Values (.csv) format. To export the data, go to "File → Export → Export data dialog box will appear → Change the file name and default directory to one's own folder →Apply to all →Export" (Fig. 8.26). Open the file from one's folder. An Excel file (in .csv format) containing the wavenumber (cm^{-1}) and absorbance (A) will appeared. Next, open the Excel file and copy the wavenumber and absorbance of selected region and paste it to the new book in origin software. Figure 8.27 shows the sample of data for PEO/PMMA blend in ClO_4^- (wavenumber: 650–600 cm^{-1}) region. Plot the graph using line. Then, reverse the scale of the graph to make it in the same way as the original FTIR spectra as shown in Figure 8.28.

To deconvolute the bands (refer Fig. 8.29), go to "Analysis → Bands and Baseline → Multiple Band Fit → Open Dialog". A dialog box (Spectroscopy: fitbands) will pop up. From there, change the band type to Gauss, Lorentz, or Voigt. The typical band type used to do deconvolution is Gauss, but still it depends on the suitability with the band shape. Then, change the number of bands. This number of bands is depending on the graph and it can up to 10 bands to deconvolute. So, before selecting the number of bands to deconvolute, one must understand what one wants to find from the unresolved bands. In this case, the purpose of doing deconvolution of ClO_4^- bands is to estimate the fraction (in ratio) of free (616 cm^{-1}) and bound (630 cm^{-1}) ClO$_4^-$ in a sample. After finish the deconvolution, click the workbook and go to the "Multiple Bands Fit Report." Extract the data (Area, Center, Width, and Height) and paste it into a new workbook in Excel (see Fig. 8.30).

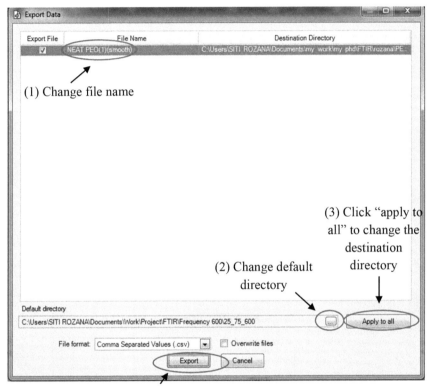

FIGURE 8.26 Step on exporting FTIR data from SPECTRUM to Excel file.

The LiClO$_4$ is highly electropositive salt and easy to dissociate to perchlorate (ClO$_4^-$) anion. The absorption band of this anion is easily detected by FTIR at range of 650–600 cm^{-1} as shown in the Figure 8.31. The overlap bands consist of two types of ClO$_4^-$, which one less and more associated of ClO$_4^-$. The less associated ClO$_4^-$ is represented band by the 616 cm^{-1} band, while for more associated, the band is at 630 cm^{-1}. In the case of PEO/PMMA blend composition with increase salt content, it is observed that the dissociation of the LiClO$_4$ is more than ~90%, suggesting that LiClO$_4$ is highly soluble in both PEO and PMMA (see Table 8.8).

230　　　　　　　　　　　　　　　Functional Polymeric Composites: Macro to Nanoscales

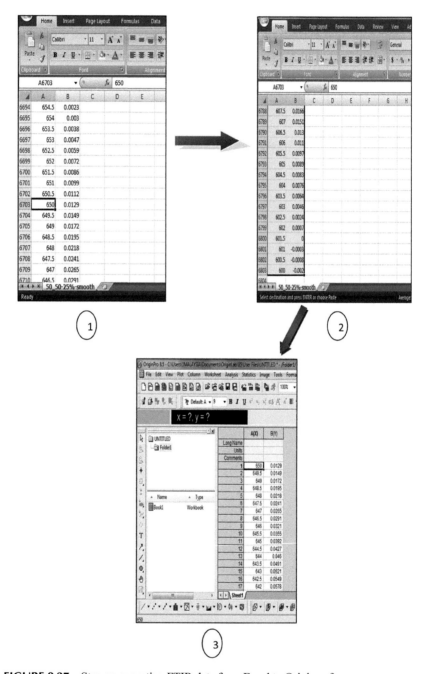

FIGURE 8.27　Step on exporting FTIR data from Excel to Origin software.

Qualitative Fourier Transform Infrared Spectroscopic Analysis 231

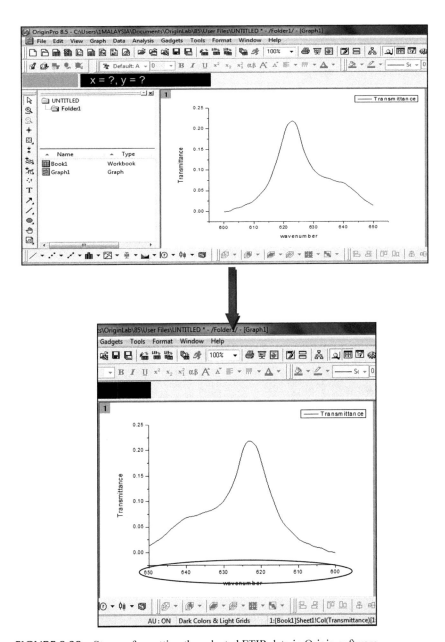

FIGURE 8.28 Step on formatting the selected FTIR data in Origin software.

232 Functional Polymeric Composites: Macro to Nanoscales

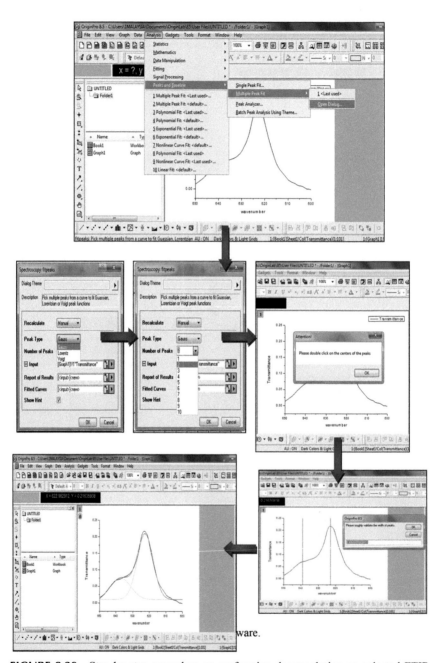

FIGURE 8.29 Step-by-step procedure on performing deconvolution on selected FTIR data in Origin software.

Qualitative Fourier Transform Infrared Spectroscopic Analysis

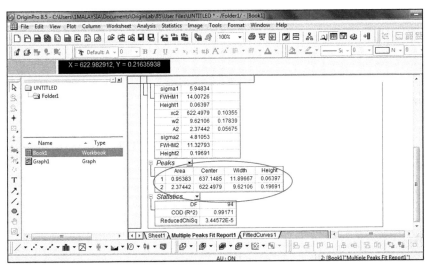

FIGURE 8.30 Extracting data from Origin software.

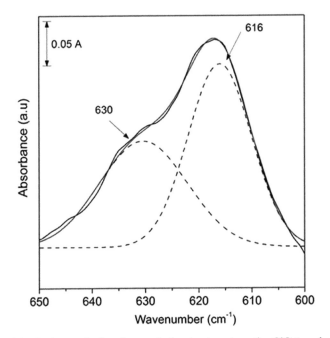

FIGURE 8.31 Before and after deconvolution treatment on the ClO_4^- peak. The solid black curve represents the original curve, meanwhile the blue dash curve represent peak after deconvolution.

TABLE 8.8 Compilation of Data after Deconvolution Technique for PEO/PMMA Blend with Addition of LiClO$_4$.

Y		ClO_4^-		$C=O$	
		f_f (%)	f_b (%)	f_f (%)	f_b (%)
PEO	0.02	96.1	3.9	–	–
	0.05	95.6	4.4	–	–
	0.07	97.4	2.6	–	–
	0.10	97.0	3.0	–	–
	0.12	97.0	3.0	–	.
	0.15	96.9	3.1	–	–
	0.20	97.5	2.5	–	–
	0.25	98.0	2.0	–	–
PMMA	0.02	92.1	7.9	71.8	28.2
	0.05	97.3	2.7	61.7	38.3
	0.07	92.9	7.1	64.9	35.1
	0.1	97.2	2.8	61.4	38.6
	0.12	53.7	46.3	56.9	43.1
	0.15	91.5	8.5	58.9	41.1
	0.25	90.1	9.9	62.0	38.0
	0.3	90.3	9.7	62.2	37.8
PEO/ PMMA 75/25	0.02	95.4	4.6	77.8	22.2
	0.05	93.5	6.5	85.0	15.0
	0.07	94.9	5.1	65.5	34.5
	0.10	93.2	6.8	44.4	55.6
	0.12	90.8	9.2	77.2	22.8
	0.15	90.8	9.2	63.4	36.6
	0.25	89.0	11.0	62.3	37.7
	0.30	88.8	11.2	79.9	20.1
PEO/ PMMA 25/75	0.02	92.0	8.0	78.8	21.2
	0.05	90.0	10.0	79.9	20.1
	0.07	91.0	9.0	77.7	22.3
	0.10	94.8	5.2	71.8	28.2
	0.12	92.7	7.3	65.3	34.7
	0.15	92.0	8.0	52.7	47.3
	0.20	81.5	18.5	50.0	50.0
	0.25	73.0	27.0	28.9	71.1
	0.30	90.0	10.0	47.6	52.4

Other than ClO_4^-, the carbonyl region is another preferred region to do the deconvolution. It is because of the interaction C=O band the highly electronegative oxygen and the cation from the salt (see Fig. 8.32). Addition of high concentration of salt causes the single band at ~1724 cm^{-1} is reduces its intensity and new band observed at lower wavenumber (~1653 cm^{-1}) is noted, depicting that more Li$^+$ is interacting with the oxygen. The variation of the free and bound C=O at different salt concentration are shown in Figure 8.33, meanwhile Table 8.8 depicts the compilation of data for both ClO_4^- and C=O after deconvolution.

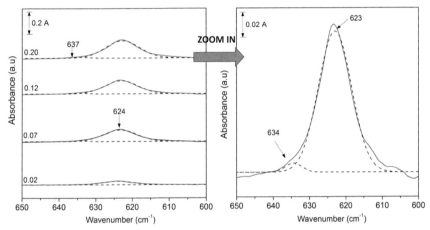

FIGURE 8.32 The before and after deconvolution treatment on the ClO_4^- band at different concentration for PEO/LiClO$_4$ system. The 623 and 634 cm^{-1} represent the free and bonded ClO_4^-, respectively.

From the table, it is observed that the fraction for the less associated ClO_4^- for all PEO/PMMA blend is increase at all LiClO$_4$ concentration. High dissociation of salt in the polymers increase the amount of Li$^+$ ion in the polymer chain area which can promote the improvement in ion conductivity of the system. However, in this complex blend where the T_g for both polymers are largely different, it is important to study in which functional groups will the salt preferred. Analysis in other area of spectrum for instance the C=O in PMMA will give more understanding the mechanism of the interaction between salt and polymer. As seen in the table, it shows that addition of salt does affect the C=O by forming another interaction between the oxygen and Li$^+$ ion. This strong interaction

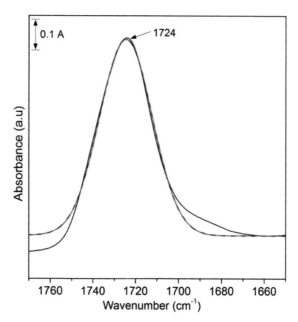

FIGURE 8.33 Before and after deconvolution treatment on the C=O peak. The solid black curve represents the original curve, meanwhile the red curve represent peak after deconvolution.

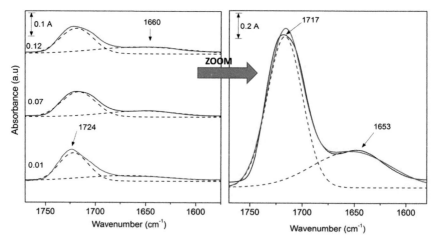

FIGURE 8.34 The before and after deconvolution treatment on the C=O band at different concentration for PEO/LiClO$_4$ system. The 1717 and 1653 cm^{-1} represent the free and bonded C=O, respectively.

causing the free C=O to be slightly decrease and at the same time forming new band at lower wavenumber. The effect is much obvious when the content of PMMA is at majority (see Table 8.8). Despite lots of discussion confirming that LiClO4 is more attracted to the PEO, from here, the study suggests that $LiClO_4$ has the possibilities to form interaction in both PEO and PMMA (Fig. 8.34).

8.5 SUMMARY

In this chapter, the qualitative analysis of polyether based polymer blends using FTIR has been discussed. The chapter begins with the introduction of solid polymer electrolyte (SPEs) and short discussion on the thermal properties of PEO/polyacrylate systems. After that, the qualitative analysis based on the interaction of PEO and PMMA with the $LiClO_4$ has been discussed in detail including the possible mechanism of that interaction. Next, the qualitative analysis based on the absorbance band has been show. The method used to extract the data is by deconvolution. The step-by step method on how to perform deconvolution is also illustrate before the discussion on the sample of work based on the PEO/PMMA with and without addition of $LiClO_4$ for perchlorate and carbonyl region.

KEYWORDS

- **solid polymer electrolytes**
- **spectroscopic analysis**
- **polyether**

REFERENCES

1. Ibrahim, S.; Mohd Yasin, S. M.; Ng, M. N.; Ahmad, R.; Johan, M. R. *Solid State Commun.* **2012,** *152,* 426–434.
2. Scrosati, B. *Application of Electroactive Polymers;* Chapman & Hall: London, 1993.
3. Gray, F. M. *Solid Polymer Electrolytes-Fundamentals and Technological Applications;* VCH: New York, 1991.

4. Gray, F. M. *Polymer Electrolytes, RSC Materials Monograph;* The Royal Society of Chemistry: Cambridge, 1997.
5. Chan, C. H.; Kammer, H. W. *J. Appl. Polym. Sci.* **2008,** *110,* 424–432.
6. Chan, C. H.; Sim, L. H.; Kammer, H. W.; Winie, T.; Nasir, N. H. A. *Mater. Res. Innov.* **2011,** *15,* 14–17.
7. Sim, L. H.; Gan, S. N.; Chan, C. H.; Yahya, R. *Spectrochim. Acta Part A.* **2010,** *76,* 287–292.
8. Yu, X. Y.; Xiao, M.; Wang, S. J.; Zhao, Q. Q.; Meng, Y. Z. *J. Appl. Polym. Sci.* **2010,** *115,* 2718–2722.
9. Xue, C.; Meador, M. A. B.; Zhu, L.; Ge, J. J.; Cheng, S. Z. D.; Puttharanat, S., et al. *Polymer.* **2006,** *47,* 6149–6155.
10. Devaux, D.; Gle, D.; Phan, T. N. T.; Gigmes, D.; Giroud, E.; Deschamps, M.; Denoyel, R.; Bouchet, R. *Chem. Mater.* **2015,** *27,* 4682–4692.
11. Alloin, F.; D'Aprea, A.; Kissi, N. E.; Dufresne, A.; Bossard, F. *Electrochim; Acta.* **2010,** *55,* 5186–5194.
12. Sivaraman, P.; Shashidhara, K.; Thakur, A. P.; Samui, A. B.; Bhattacharyya, A. R. *Poly. Eng. Sci.* **2015,** *55,* 1536–1545.
13. Armand, M. B.; Chabagno, J. M.; Duclot, M. *Fast Ion Transport in Solids;* Vashisha, P., Mundy, J. N., Shenoy, G. K., Eds.; Elsevier North Holland: New York, 1979; pp 131–136.
14. Kim, C. S.; Oh, S. M. *Electrochim. Acta.* **2000,** *45,* 2101–2109.
15. Caruso, T.; Capoleoni, S.; Cazzanelli, E.; Agostino, R. G.; Villano, P.; Passerini, S. *Ionics.* **2002,** *8,* 36–43.
16. Rocco, A. M.; Fonseca, C. P.; Pereira, R. P. *Polymer.* **2002,** *43,* 3601–3609.
17. Floriańczyk, Z.; Marcinek, M.; Wieczorek, W.; Langwald, N. *Polish J. Chem.* **2004,** *78,* 1279–1304.
18. Appetecchi, G. B.; Croce, F.; Scosati, B. *Electrochim. Acta.* **1995,** *40,* 991–997.
19. Bohnke, O.; Frand, G.; Rezrazi, M.; Rousselot, C.; Truche, C. *Solid State Ionics.* **1993a,** *66,* 97–104.
20. Bohnke, O.; Frand, G.; Rezrazi, M.; Rousselot, C.; Truche, C. *Solid State Ionics.* **1993b,** *66,* 105–112.
21. Song, J. Y.; Wang, Y. Y.; Wan, C. C. *J. Power Sources.* **1999,** *77,* 183–197.
22. Vondrak, J.; Reiter, J.; Velická, J.; Sedlaříková, M. *Solid State Ionics.* **2004,** *170,* 79–82.
23. Uma, T.; Mahalingam, T.; Stimming, U. *Mater. Chem. Phys.* **2005,** *90,* 245–249.
24. Tsuchida, E.; Ohno, H.; Tsunemi, K. *Electrochim. Acta.* **1983,** *28,* 591–595.
25. Choe, H. S.; Giaccai, J.; Alamgir, M.; Abraham, K. M. *Electrochim.a Acta.* **1995,** *40,* 2289–2293.
26. Jacob, M. M. E.; Prabaharan, S. R. S.; Radhakrishna, S. *Solid State Ionics.* **1997,** *104,* 267–276.
27. Han, H-S.; Kang, H.-R.; Kim, S.-W.; Kim, H.-T. J. Power Sources. **2002,** *112,* 461–468.
28. Abraham, K. M.; Alamgir, M. J. *Electrochem. Soc.* **1990,** *137,* 1657–1658.
29. Labreche, C.; Levesque, I.; Prud'homme, J. *Macromolecules.* **1996,** *29,* 7795–7801.
30. Hamon, L.; Grohens, Y.; Soldera, A.; Holl, Y. *Polymer.* **2001,** *42,* 9697–9703.
31. Rocco, A. M.; Fonseca C. P.; Loureiro, F. A. M.; Pereira, R. P. *Solid State Ionics.* **2004,** *166,* 115–126.

Qualitative Fourier Transform Infrared Spectroscopic Analysis 239

32. Mahendran, O.; Chen, S. Y.; Chen-Yang, Y. W.; Lee, J. Y.; Rajendran, S. *Ionics.* **2005,** *11,* 251–258.
33. Chan, C. H.; Kammer, H. W. *Ionics.* **2015,** *21,* 927–934.
34. Chan, C. H.; Kammer, H. W.; Sim, L. H.; Mohd Yusoff, S. N. H.; Hashifudin, A.; Winie, T. *Ionics.* **2014,** *20,* 189–199.
35. Abd Karim, S. R.; Sim, L. H.; Chan, C. H.; Ramli, H.; *Macromol. Symp.* **2015,** *354,* 374–383.
36. Murata, K.; Izuchi, S.; Yoshihisa, Y. *Electrochim. Acta.* **2000,** *45,* 1501–1508.
37. Othman, L.; Chew, K. W.; Osman, Z. *Ionics.* **2007,** *13,* 337–342.
38. Berthier, C.; Gorecki, W.; Minier, M.; Armand, M. B.; Chabagno, J. M.; Rigaud, P. *Solid State Ionics.* **1983,** *11,* 91–95
39. Croce, F.; Brown, S. D.; Greenbaum, S. G.; Slane, S. M.; Salomon, M. *Chem. Mater.* **1993,** *5,* 1268–1272.
40. Stallworth, P. E.; Greenbaum, S. G.; Croce, F.; Slane, S.; Salomon, M. *Electrochimica. Acta.* **1995,** *40,* 2137–2141.
41. Gupta, R. K.; Rhee, H. W. *Electrochim. Acta.* **2012,** *76,* 159–164.
42. Rajendran, S.; Mahendran, O.; Kannan, R. *Solid State Electrochem.* **2002c,** *6,* 560–564.
43. Tan, S. M.; Johan, M. R. *Ionics.* **2011,** *17,* 485–490.
44. Pedrosa, P.; Pomposo, J. A.; Calahorra, E.; Cortazar, M. *Polymer.* **1995,** *36,* 3889–3897.
45. Zheng, H.; Zheng, S.; Guo, Q. *J. Polym. Sci. Polym. Chem.* **1989,** *35,* 3169–3179.
46. Sotele, J. J.; Soldi, V.; Pires, A. T. N. *Polymer.* **1997,** *38,* 1179–1185.
47. Rocco A. M.; Pereira, R. P.; Felisberti, M. I. *Polymer.* **2001,** *42,* 5199–5205.
48. Silvestre, C.; Cimmino, S.; Martuscelli, E.; Karasz, F. E.; Macknight, W. J. *Polymer.* **1987,** *28,* 1190–1199.
49. Martuscelli, E.; Pracella, M.; Yue, W. P. Polymer, **1984,** *25,* 1097–1106.
50. Ramana Rao, G.; Castiglioni, C.; Gussoni, M.; Zerbi, G.; Martuscelli, E. *Polymer.* **1985,** *26,* 811–820.
51. Shafee, E. E.; Ueda, W. *Eur. Polym. J.* **2002,** *38,* 1327–1335.
52. Li, X.; Hsu, S. L. *J. Polym. Sci. Polym. Phys. Ed.* **1984,** *22,* 1331–1342.
53. Martuscelli, E.; Silvestre, C.; Addonizio, M. L.; Amelino, L. *Makromol. Chem.* **1986,** *187,* 1557–1571.
54. Talibuddin, S.; Wu, L.; Runt, J.; Lin, J. S. *Macromolecules.* **1996,** *29,* 7527–7535.
55. Martuscelli, E.; Marchena, C.; Nicolais, L. *Future Trends in Polymer Science and Technology – Polymers: Commodities or Specialties;* Technomic Pub. Co.: Basel, Switzerland,1987; p 247.
56. Schantz, S. *Macromolecules.* **1997,** *30,* 1419–1425.
57. Woo, E. M.; Mandal, T. K.; Chang, L. L.; Lee, S. C. *Polymer.* **2000,** *41,* 6663–6670.
58. Mandal, T. K.; Kuo, J. F.; Woo, E. M. *J. Polym. Sci. Polym. Phys. Ed.* **2000,** *38,* 562–572.
59. Pfefferkorn, D.; Kyeremateng, S. O.; Busse, K.; Kammer, H. W.; Thurn-Albrecht, T.; Kressler, J. *Macromolecules.* **2011,** *44,* 2953–2963.
60. Katime, I.; Cadenato, A. *Mater. Lett.* **1995,** *22,* 303–308.
61. Pereira, R. P.; Rocco, A. M. *Polymer.* **2005,** *46,* 12493–12502.
62. Mu, D.; Huang, X. R.; Lu, Z. Y.; Sun, C. C. *Macromolecules.* **2012,** *45,* 2035–2049.

63. Lü, H.; Zheng, S.; Tian, G. *Polymer.* **2004,** *45,* 2897–2909.
64. Zhong, Z.; Guo, Q. *Polymer.* **1998,** *39,* 517–523.
65. Hsieh, K. H.; Ho, K. S.; Wang, Y. Z.; Ko, S. D.; Fu, S. C. *Synth. Met.* **2001,** *123,* 217–224.
66. Baskaran, R.; Selvasekarapandian, S.; Kuwata, N.; Kawamura, J.; Hattori, T. *Mater. Chem. Phys.* **2006,** *98,* 55–61.
67. Wang, M.; Braun, H. G.; Meyer, E. *Polymer.* **2003,** *44,* 5015–5021.
68. Ito, H.; Russell, T. P.; Wignall, G. D. *Macromolecules.* **1987,** *20,* 2213–2220.
69. Guo, Q.; Harrats, C.; Groeninckx, G.; Koch, M. H. Z. *Polymer.* **2001,** *42,* 4127–4140.
70. Ramesh, S.; Koa,y H. L.; Kumutha, K.; Arof A. K. *Spectrochim. Acta Part A.* **2007,** *66,* 1237–1242.
71. Martuscelli, E.; Demma, G.; Rossi, E.; Segre, A. L. *Polym. Commun.* **1983a,** *24,* 266–267.
72. Martuscelli, E.; Silvestre, C.; Bianchi, L. *Polymer.* **1983b,** *24,* 1458–1468.
73. Parizel, N.; Laupretre, F.; Monnerie, L. *Polymer.* **1997,** *38,* 3719–3725.
74. Straka, J.; Schmidt, P.; Dybal, J.; Schneider, B.; Spěváček, J. *Polymer.* **1995,** *36,* 1147–1155.
75. Marco, C.; Fatou, J. G.; Gomez, M. A.; Tanaka, H.; Tonelli, A. E. *Macromolecules.* **1990,** *23,* 2183–2188.
76. Lartigue, C.; Guillermo, A.; Cohen-Addad, J. P. *J. Polym. Sci. B: Polym. Phys.* **1997,** *35,* 1095–1105.
77. Nawawi, M. A.; Sim, L. H.; Chan, C. H. *Int. J. Chem. Eng. Appl.* **2012,** *3,* 410.
78. Hashifudin, A.; Sim, L. H.; Chan, C. H.; Kammer, H. W.; Yusoff, S. N. M. Y. *Polym. Res. J.* **2012,** *7,* 196.
79. Cimmino, S.; Martuscelli, E.; Silvestre, C. *Makromol. Chem.* **1990,** *191,* 2447–2454.
80. Elberanchi, A.; Daro, A.; David, C. *Eur. Polym. J.* **1999,** *35,* 1217–1228.
81. Jeddi, K.; Qazvini, N. T.; Jafari, S. H.; Khonakdar, H. A. *J. Polym. Sci. Part B: Polym. Phys.* **2010,** *48,* 2065–2071.
82. Utracki, L.A. Polymer Alloys and Blends: Thermodynamics and Rheology. Munich: Hanser GmbH and Company, 1989.
83. Olabisi, O.; Robeson, L. M.; Shaw, T. Polymer-polymer Miscibility. New York: Academic Press, 1979.
84. Gordon, M.; Taylor, J. S.; *J. Appl. Chem.* **1952,** *2,* 493–500.
85. Lu, X.; Weiss, R. A. *Macromolecules.* **1992,** *25,* 3242–3246.
86. Lodge, T. P.; Wood, E. R.; Haley, J. C. *J. Polym. Sci. Part B: Polym. Phys.* **2006,** *44,* 756.
87. Fernandes, A. C.; Barlow, J. W.; Paul, D. R. *J. Appl. Polym. Sci.* **1986,** *32,* 5481.
88. Rodgers, P. A.; Paul, D. R.; Barlow, J. W. *Macromolecules.* **1991,** *24,* 4101.
89. Nasir, N. H. A.; Chan, C. H.; Kammer, H. W.; Sim, L. H.; Yahya, M. Z. A. *Macromol. Symp.* **2010,** *290,* 46.
90. Smith, B. C. *Fundamentals of Fourier Transform Infrared Spectroscopy;* CRC Press: Florida, 1996.
91. Rajendran, S.; Kannan, R.; Mahendran, O. *J. Power Sour.* 2001, *96,* 406–410.
92. Ghelichi, M.; Qazvini, N. T.; Jafari, S. H.; Khonakdar, H. A.; Farajollahi, Y.; Scheffler, C. *J. Appl. Polym. Sci.* **2013,** *129,* 1868–1874.
93. Rajendran, S.; Sivakumar, M.; Subadevi, R. *Solid State Ionics.* **2004,** *167,* 335–339.

Qualitative Fourier Transform Infrared Spectroscopic Analysis

94. Sivakumar, M.; Subadevi, R.; Rajendran, S.; Wu, N. L.; Lee, J. Y. S*olid State Ionics.* **2006,** *97,* 330–336.
95. Rajendran, S.; Uma, T. *Mater. Lett.* **2000,** *45,* 191–196.
96. Meneghetti, P.; Webber, A.; Qutubuddin, S. *Electrochim. Acta.* **2004,** *49,* 4923–4931.
97. Deka, M.; Kumar, A. *Electrochim. Acta.* **2010,** *55,* 1836–1842.
98. Mahendran, O.; Rajendran, S. *Ionics.* **2003,** *9,* 282–288.
99. Rajendran, S.; Mahendran, O.; Kannan, R. *J. Phys. Chem. Solids.* **2002,** *63,* 303–307.
100. Hoffmann, C. L.; Rabolt, J. F. *Macromolecules.* **1996,** *29,* 2543–2547.
101. Cipraci, D.; Jacob, K.; Tannenbaum, R. *Macromolecules.* **2006,** 39, 6565–6573.
102. Ramesh, S.; Liew, C.W. *Measurement.* **2013,** *46,* 1650–1656.
103. Cesteros, L. C.; Meaurio, E.; Katime, I. *Macromolecules.* **1993,** *26,* 2323–2330.
104. Kauppinen, J. K.; Moffatt, D. J.; Mantsch, H. H; Cameron, D. G. *Appl. Spectrosc.* **1981,** *35,* 271–276.
105. Salim, N. V. *Nanostructured Block Copolymer Blends and Complexes via Hydrogen Bonding Interactions.* PhD Thesis, 2009.

INDEX

A

Acetobacter xylinum, 164
1-(4-Acryloyloxybutyl)-3-methylimidazolium bromide (AcMIMBr), 38
AcVIMBr, *in situ* copolymerization of AcMIMBr with, 39–40
Aleurites fordii, 133
1-Allyl-3-methylimidazolium bromide (AMIMBr), 34
Amines vapors detection by polymerized tung oil-based quartz crystal microbalance sensor, 144–145
experimental
materials and sensor fabrication, 146–148
set up, 148–149
results and discussions, 149–152
Amorphous polymers, 7

B

Bacterial cellulose (BC), 164
Band absorbance, qualitative analysis, 225–237. *See also* Fourier transform infrared (FTIR) spectroscopic method
deconvolution of, 225–226
step-by-step guidelines, 226–237
Bio-based phenolic resin production, 77
Biomass feedstock for production of phenolic resin, 83–86
Bionanocomposites, 179
1-Butyl-3-methylimidazolium chloride (BMIMCl), 36

C

Carr's index, 62
Castor oil, 132–133
Cellulose, 34
chemical structures of, 35

defibrillation of, 165
Cellulose microfibrils (MCF), 165
Cellulose nanocrystals (CNC), 164, 167
acidic hydrolysis, 170–171
alkali treatment, 169
delignification, 169–170
mechanical and physical properties, 171–172
TEM image, 168
Cellulose nanofibrils (CNF), 164, 172
physical disintegration, 175
Gaulin homogenizer, 176
grinding, 176–177
high speed blender, 177
micofluidizer, 176
ultrasonication, 177–178
pretreatment, 173
acidified chlorite, 174–175
enzymatic hydrolysis, 175
TEMPO-mediated oxidation, 174
Cellulose-poly(AcMIMBr-*co*-AcVIMBr) composite material, preparation, 40
Cellulose-poly(AcMIMBr) composite material, preparation, 39
Cellulose-polymeric ionic liquid composite, 34
materials preparation, 38–43
Cellulose-poly(VVBnIMCl) composite material preparation, 43
Chemiresistors. *See* Metal oxide sensors
Chitin
chemical structures of, 35
dispersion and nanofiber film by regeneration from chitin ion gel with AMIMBr, 45
dissolution and gelation, with AMIMBr, 44

244   Functional Polymeric Composites: Macro to Nanoscales

nanofibrous composite materials using ionic liquid, preparation of, 43–48
synthetic polymer composite materials by physical and chemical approaches, 46
top-down and bottom-up approaches for production of, 38
Chitin nanofiber-poly(vinyl alcohol) composite film, 34
procedure for preparation of, 46–47
Chloroprene rubber (CR), 7
Colloidal MCC, 58–59
Commercial MCC products, 60–61
Compatibilizer, 5
Compatible blends, 5
Copolymerization, 38
Cotton seed oil, 133
Cross-linked polymers from oils, 134–135
Crystallite gel model, 65

D

Debye relaxation exponent estimation, 121–126
Deconvolution treatment
before and after, 235–237
compilation of data, 234
data from Excel, 230
data from SPECTRUM, 229
extracting data from Origin software, 233
step-by-step procedure, 232
step on formatting, 231
steps, 226–228
Differential scanning calorimetry (DSC) thermogram, 201
Dimethylol phenol and trimethylol phenol structure, 81–82

E

Ecballium elaterium, 136
Elastomers, 6–7
Empty fruit bunch (EFB) fibers, 91
Epoxidized natural rubber (ENR), 7

N-Ethylpyridinium chloride, 35–36
Extrusion-spheronization process, 65–67

F

Fast pyrolysis, 85–88. *See also* Phenolic compounds production, from biomass
Formaldehyde to phenol, addition and condensation reaction, 82–83
Fourier transform infrared (FTIR) spectroscopic method, 198
qualitative analysis of, 207
assignment of absorption band, 208–210
attenuated total reflectance base assembly, schematic diagram of, 208
band absorbance, 225–237
characteristic bands, 212–214
PEO/LICLO system, 214–216
PEO, PMMA, and LICLO, 211–214
PMMA/LICLO system, 217–218
rules, 210–211
salt-free and salt-added peo/pmma blend, 218–225

G

Gas sensor, 138–139. *See also* Sensor
Gordon-Taylor Equation, 202–203
Gossypium malvaceae, 133
Guaiacol, chemical structure, 85

H

Hardwood, typical proximate composition, 86
Hausner ratio values, 62
Helianthus annuus, 134

I

Imidazolium-type ionic liquids, 34
Immiscible polymer blends, applications, 4–5
Impedance spectroscopy (IS), 98
In situ polymerization method, 38–43
Interpenetrating polymer network (IPN) system, 38

Index 245

Ionic conductivity, 98
Isotactic poly(styrene) (*i*PS)/PS, 3–4

L

Linseed oil, 133
  reaction mechanism of polymerization, 134–135
Liquefaction of lignocellulosic biomass, 90–92

M

Mechanical properties, 9
Metal oxide sensors, 139. *See also* Sensor
Microcrystalline cellulose (MCC)
  functionality of, 56–57
  manufactured by purification and partial depolymerization, 56
  structure and manufacture
    available forms, 57–59
    co-processed blend, 58–59
    inter-grade variability in, 59, 62
    MCC-solvent interaction, 69–70
    models, pelletization aid functionality of, 64–65
    as pelletization aid, 64
    pharmaceutical processes, 62–64
    using non-aqueous solvents as moistening liquid, 67–69
    using water as moistening liquid, 65–67
Microfibrillated cellulose (MFC), 165
Miscible polymer blends, 3–4
Multi-component polymeric systems, classification, 8

N

Nanocellulose, 164–165
  BC, 178
  CNC, 167–172
  CNF, 172–178
  types, 167
  from various methods and sources, 166

Nanocomposites, 178–180
  bioadsorbents, 185–186
  biomedical applications, 183–184
  electronics, 184–185
  rheology modifier, 186
  solution casting
    organic solvent based, 182
    water based, 181–182
  thermal processing, 182–183
Nanofibrillated cellulose (NFC), 165
Natural oils, 130
  major fatty acid composition in, 132
Nitrile-butadiene rubber (NBR), 7
Novolac, 76
Nylon 11, 136–137
Nyquist plot
  bulk resistance determintation from
    using graphical method, 99–106
    using mathematical method, 106–116

O

Oligoesters (OE), 199
Origin® software, 98, 230–233

P

PEO/polyacrylate systems, thermal properties and spectroscopic analysis, 199–207
  spherulitic morphology, 200
Phenol formaldehyde (PF)
  types, 76
Phenolic compounds production, from biomass
  fast pyrolysis, 85–88
  liquefaction, 90–92
  vacuum pyrolysis, 88–89
Plant oils, 130–131
Poly(benzyl methacrylate) (PBzMA), 200
Poly(β-hydroxybutyrate)/poly(*p*-vinylphenol), 3–4
Poly(butyl methacrylate) (PBMA), 200

Poly(cyclohexyl methacrylate) (PCMA), 200

Poly(ε-caprolactone) (PCL)/ poly(styrene-*co*-acrylonitrile), 5

Polyester resins (PER), 199

Poly(ethylene oxide) (PEO)/poly(methyl methacrylate) (PMMA), 3–4

Poly(iso-butyl methacrylate) (PiBMA), 200

Polymer blending, 199–200

Polymer-clay composites, 5

Polymerized linseed oil based quartz crystal microbalance sensor and its application, 152

experimental, 153

result and discussion, 153–156

Polymerized oil applications, 135–137

Polymers, classification, 6–8

Poly(methyl vinyl ether-maleic acid) (PMVE-MAc), 199

Poly(n-butyl methacrylate) (PnBMA), 200

Poly(phenyl methacrylate) (PPhMA), 200

Poly(propyl methacrylate (PPMA), 200

Polysaccharide-based composite materials, 36

Poly(styrene) (PS)/poly(phenylene oxide) (PPO), 3–4

Poly(tert-butyl methacrylate) (PtBMA), 200

Poly(vinyl alcohol) (PVA), 199

Poly(vinyl alcohol) (PVA)/poly(vinyl pyrrolidone) (PVP), 5

Poly(vinylidene fluoride) (PVDF)/ PMMA, 5

Poly(4-vinylphenol-*co*-2-hydroxyethyl methacrylate) (PVPh-HEM), 200

Poly(vinyl phenol) (PVPh), 199

Powdered MCC, 58

PP-grafted-GMA (PP-*g*-GMA) compatibilized PC/PP system, 21

PP/NR blends, 5–6

Q

Quartz crystal microbalance (QCM) sensor. *See also* Sensor

applications of, 141–143

Sauerbrey equation for, 140

schematic diagram of experimental setup, 143

Quinone methide, 81

R

Resin and its chemistry, 77

novolac resin, 78, 80

resol-type PF resin, 80–83

Resol, 76

addition and condensation steps for, 82

Ricinus communis, 132–133

S

Semicrystalline/amorphous polymer blends, 3–5

mechanical properties, 9–23

Semicrystalline polymers, 7

Sensor, 137

classification of, 138

gas sensor, 138–139

metal oxide sensors, 139

quartz crystal microbalance (QCM) sensor, 140–143

surface acoustic wave sensor, 139–140

Softwood, typical proximate composition, 86

Soja hispida, 133

Solid polymer electrolytes (SPES), 198–199

Soybean oil, 133

Spheronization, 66–67

Sponge model, 65

Stable CNC, 164

Styrene-butadiene rubber (SBR), 7

Sunflower oil, 134

Surface acoustic wave sensor, 139–140

Index

Surface-initiated ring-opening copolymerization of LA/CL from chitin nanofiber film, 48
Syringol, chemical structure, 85

T

2,2,6,6-Tetramethylpiperidine-1-oxyl radical (TEMPO), 165
Thermoplastics, 6
Thermosets, 6–7
Thermosetting resins, 76

Triglycerides oils, chemical structure, 131
Tung oil, 133

V

Vacuum pyrolysis, 88–89. *See also* Phenolic compounds production, from biomass

W

Wet massing (granulation) behaviors, 63

PGSTL 11/17/2017